U.S. Department of Justice
Office of Justice Programs
Office for Victims of Crime

U.S. Department of Justice
Office of Justice Programs
810 Seventh Street NW.
Washington, DC 20531

Eric H. Holder, Jr.
Attorney General

Laurie O. Robinson
Assistant Attorney General

Joye E. Frost
Acting Director, Office for Victims of Crime

Office of Justice Programs
Innovation • Partnerships • Safer Neighborhoods
www.ojp.usdoj.gov

Office for Victims of Crime
www.ovc.gov

NCJ 231171

Preparation of this publication was supported by grant number 2005–VF–GX–4001, awarded by the Office for Victims of Crime, Office of Justice Programs, U.S. Department of Justice. The opinions, findings, and conclusions or recommendations expressed in this document are those of the author and do not necessarily represent the official position or policies of the U.S. Department of Justice.

The Office of Justice Programs (OJP), headed by Assistant Attorney General Laurie O. Robinson, provides federal leadership in developing the Nation's capacity to prevent and control crime, administer justice, and assist victims. OJP has seven components: the Bureau of Justice Assistance; the Bureau of Justice Statistics; the National Institute of Justice; the Office of Juvenile Justice and Delinquency Prevention; the Office for Victims of Crime; the Community Capacity Development Office; and the Office of Sex Offender Sentencing, Monitoring, Apprehending, Registering, and Tracking. More information about OJP can be found at http://www.ojp.gov.

FIRST RESPONSE TO VICTIMS OF CRIME

A GUIDEBOOK FOR LAW ENFORCEMENT OFFICERS

National Sheriffs' Association

July 2010

NCJ 231171

ACKNOWLEDGMENTS

The Office for Victims of Crime (OVC) wishes to acknowledge the National Sheriffs' Association (NSA) for initiating development of this guidebook and Timothy O. Woods, J.D., M.A., LL.M., Director of Research, Development & Grants at NSA, for writing it.

Recognition is also due to the many individuals who generously contributed their professional expertise and personal experiences in support of this project—or the earlier *First Response to Victims of Crime* handbooks on which this project builds—and without whose encouragement and extensive support the project could never have been undertaken, much less completed. Accordingly, OVC and NSA extend special thanks to the following persons: Nora J. Baladerian, Ph.D., Disability, Abuse and Personal Rights Project; Imelda Buncab, Coalition to Abolish Slavery & Trafficking; Florrie Burke, Anti-Trafficking Program, Safe Horizon; Gail Burns-Smith, Connecticut Sexual Assault Crisis Services; Betsy Cantrell, Consultant; Nancy Chandler, National Children's Alliance; Marc P. Charmatz, J.D., National Association of the Deaf Law and Advocacy Center; Doreen M. Croser, American Association on Intellectual and Developmental Disabilities; Leigh Ann Davis, The Arc of the United States; Marcie H. Deitch, formerly International Association of Chiefs of Police; Sharon D'Eusanio, Division of Victim Services and Criminal Justice Programs, Office of the Florida Attorney General; Robin F. Finegan, Finegan, Flannigan & Associates; Krista R. Flannigan, J.D., Finegan, Flannigan & Associates; Maria Jose Fletcher, J.D., Florida Immigrant Advocacy Center; Stephanie Frogge, formerly Mothers Against Drunk Driving, National Office; Xu Gao, LL.M., Law Office of Susan Gao; Laura Germino, Coalition of Immokalee Workers; Brian Hance, formerly Alzheimer's Association; Candace J. Heisler, J.D., Heisler and Associates; Terri J. Hicks, NSA; Ronald S. Honberg, J.D., National Alliance on Mental Illness; Catherine H. Hoog, Abused Deaf Women's Advocacy Services; Jeri Houchins, Back to Life; Donna M. Hughes, Ph.D., Women's Studies Program, University of Rhode Island; Ann Hutchison, formerly Disaster/Terrorism Center for Victim Assistance, Texas Office of

First Response to Victims of Crime

Emergency Management; Ann Jordan, J.D., Initiative Against Trafficking in Persons, Global Rights; Angela M. Kaufman, Department on Disability, City of Los Angeles; Alan Ping-Lun Lai, Crime Victims Program, Chinese Information and Service Center; Gerald Landsberg, D.S.W., School of Social Work, New York University; Linda E. Ledray, Ph.D., Sexual Assault Resource Service; Gail London, Federal Law Enforcement Training Center; Margaret MacDonnell, Migration & Refugee Services, U.S. Conference of Catholic Bishops; Candace Matthews, English for Academic Purposes Program, The George Washington University; Christine A. Mayman, formerly Court Appointed Special Advocates of the Eastern Panhandle; Chuck McCormick, Federal Bureau of Investigation (retired); Nicole McGee, formerly National MultiCultural Institute; Nyssa Mestas, Migration & Refugee Services, U.S. Conference of Catholic Bishops; Heather Moore, Coalition to Abolish Slavery & Trafficking; Lisa Nerenberg, Consultant; Steve Otto, formerly Civilian Police Programs, U.S. Department of State; Catherine Pierce, Office to Monitor and Combat Trafficking in Persons, U.S. Department of State; Peggie Reyna, Peace Over Violence; Nancy Ruhe, National Organization of Parents Of Murdered Children; Sgt. Doreen Russo, formerly Sexual Battery and Child Abuse Unit, City of Miami Police Department; Elizabeth Salett, National MultiCultural Institute; Marilyn J. Smith, Abused Deaf Women's Advocacy Services; Vickie Smith, National Center on Domestic and Sexual Violence; Deborah Spungen, Anti-Violence Partnership of Philadelphia; Kavitha Sreeharsha, J.D., Asian Pacific Islander Legal Outreach; John H. Stein, J.D., formerly National Organization for Victim Assistance; D.J. Stemmler, Center for Assistive Technology, University of Pittsburgh Medical Center; Randolph Thomas, formerly Criminal Justice Academy, South Carolina Department of Public Safety; Deborah D. Tucker, National Center on Domestic and Sexual Violence; Nancy Turner, Law Enforcement Leadership Initiative on Violence Against Women, International Association of Chiefs of Police; Cheryl G. Tyiska, formerly National Organization for Victim Assistance; Victor Vieth, J.D., formerly National Center for the Prosecution of Child Abuse, American Prosecutors Research Institute; Christina Walsh, National Center on Domestic and Sexual Violence; Juliet Walters, formerly National Center on Domestic and Sexual Violence; Kimberly Wisseman, Disability Services, SafePlace; James A. Wright, formerly NSA; and Mary T. Zdanowicz, J.D., formerly Treatment Advocacy Center.

NSA also wishes to especially acknowledge and express its appreciation to Meg Morrow, of OVC, for her guidance, leadership, and forbearance in shepherding this project to completion.

Finally, gratitude is extended to the many victim services and law enforcement practitioners who used the earlier *First Response to Victims of Crime* handbooks and called or wrote the author and OVC with their recommendations for improvements or enhancements to any future publications.

CONTENTS

Message From the Director ix

I. **Basic Guidelines on First Response to Victims of Crime** 1
   General Tips on Responding to Victims' Three Major Needs 2

II. **First Response to Individual Types of Crime Victims** 7
   Older Victims 7
   Child Victims 10
   Victims Who Have a Disability 13
   Americans with Disabilities Act of 1990 and
      Section 504 of the Rehabilitation Act of 1973 17
   Victims Who Have Alzheimer's Disease 18
   Victims Who Have a Mental Illness 21
   Victims With Mental Retardation 25
   Victims With Blindness or Vision Impairment 28
   Victims Who Are Deaf or Hard of Hearing 30
   Victims With a Disability Affecting Physical Mobility 35
   Immigrant Victims 37

III. **First Response to Specific Types of Criminal
   Victimization** 43
   Victims of Sexual Assault 43
   Victims of Domestic Violence 47
   Victims of Drunk Driving Crashes 50
   Survivors of Homicide Victims 53
   Victims of Human Trafficking 57
   Victims of Mass Casualty Crimes 65

IV. **Directory of National Service Providers** 71

V. **Endnotes** 81

MESSAGE FROM THE DIRECTOR

Whenever a crime is committed, law enforcement officers are usually the first to arrive on the scene and to interact with victims. Law enforcement officers have more contact with crime victims than any other criminal justice professional. This makes their role critical and puts them in a unique position to assist victims immediately after the crime and encourage and facilitate victim participation in the criminal justice system. The initial response to a victim will have a long-lasting impact on that individual's view of the justice system and participation in the investigation and prosecution of the crime. The first response also is a key factor in whether or not a victim ultimately accesses needed services and assistance, such as crisis intervention, counseling, financial compensation, information, referrals to community programs, and help in navigating the justice process.

In 2000, the Office for Victims of Crime (OVC) published *First Response to Victims of Crime,* a handbook for law enforcement officers to help them better understand and meet the needs of victims of crime. It offered basic guidelines for approaching and interacting with older victims, sexual assault victims, child victims, domestic violence victims, and survivors of homicide victims. In 2001, OVC published an updated handbook that included an additional section on responding to victims of alcohol-related driving crashes. Over the years, *First Response to Victims of Crime* has been one of the most requested resources produced by OVC.

In 2002, OVC released a companion handbook entitled *First Response to Victims of Crime Who Have a Disability.* With that handbook, OVC aimed to further increase the capacity of law enforcement to respond to particular populations of crime victims in a sensitive and effective manner, recognizing the unique needs of certain individuals. The handbook specifically offered guidance and tips on approaching and interacting with victims who have Alzheimer's disease, mental illness, mental retardation, or who are blind, vision impaired, deaf, or hard of hearing.

Over the past few years, new issues have emerged with the changing demographics in the United States, the occurrence of a number of high-profile mass casualty crimes, and a growing awareness of the prevalence of the crime of human trafficking. It became clear that a new, updated, expanded guidebook for law enforcement was needed. This new guidebook consolidates and updates the information in the earlier handbooks and expands the information with additional sections on responding to immigrant victims, victims with a disability affecting physical mobility, victims of human trafficking, and victims of mass casualty crimes.

In one resource, this guidebook offers valuable, user-friendly information for law enforcement on how to respond to a wide range of victims. The guidebook is not intended to be a training manual and does not claim to offer guidance on responding in every possible situation. It attempts, however, to highlight the most salient issues involved for victims of certain crimes and for certain populations of victims.

An introductory section contains general guidelines and tips, and individual sections include information on responding to victims of particular crimes and to specific populations of victims. Additionally, the guidebook includes a section on federal laws that prohibit discrimination against individuals with disabilities and a directory of resources, including organizations representing the interests of the victim populations addressed in the guidebook.

The updated guidebook is being released with a companion video entitled *First Response to Victims of Crime* which highlights and amplifies several of the topics covered in the guidebook. It will complement the guidebook and enhance its instructional messages, offering an alternative means for conveying information in the guidebook to law enforcement officers and other first responders. It accomplishes this through interviews with victims, survivors, law enforcement officers, and victim advocates. OVC hopes the guidebook and video will serve as useful resources in a variety of law enforcement training settings, such as inservice trainings, roll calls, and recertification programs.

Assistance from law enforcement makes a significant difference for victims. Victims consistently express tremendous gratitude and appreciation for the reassurance and help received from the responding officer. *First Response to Victims of Crime* serves as a reminder that all crime victims deserve to be treated with compassion, sensitivity, and respect. A response encompassing all those qualities undoubtedly will serve to increase the effectiveness of the entire criminal justice system.

SECTION I

BASIC GUIDELINES ON FIRST RESPONSE TO VICTIMS OF CRIME

The way people cope as victims of crime depends largely on their experiences and on how others treat them immediately after the crime. As a law enforcement officer, you are usually the first official to interact with victims. For this reason, you are in a unique position to help victims cope with the immediate trauma of the crime as well as to help them regain a sense of security and control over their lives.

The circumstances of a crime frequently dictate when and how responding officers first address victims and their needs. You may have to delay fully attending to victims as you juggle many tasks, such as determining what other emergency services are needed and calling for them, evacuating people from the site, securing the crime scene, or advising other public safety personnel upon their arrival. As soon as the responding officer's most urgent tasks have been completed, however, attention can be focused on victims and their needs. At that point, how you approach and relate to victims, explain your various law enforcement responsibilities, and work with victims is crucial to their recovery.

Moreover, the responding officer's awareness of the needs of victims, the many dimensions and consequences of crime for victims, common responses to victimization, and the particular needs of distinct victim populations can help the officer avoid a revictimization of victims. Conversely, inadvertently making comments or asking questions that are hurtful to victims, seemingly implying that victims are partially responsible for their own victimization, forgetting to return property taken from victims as evidence, or in any other way unknowingly being insensitive to victims can inflict a second victimization on them.

By approaching victims in a respectful and supportive manner, officers can gain their trust and cooperation. Victims may then be more willing to provide detailed information about the crime to officers and later to investigators and prosecutors, which, in turn, will lead to the conviction of more criminals. But always remember that you are there for the victim; crime victims are not just witnesses who are there to assist you with your duties. In other words, put victims first!

You can better respond to individual types of crime victims and specific types of criminal victimizations by first understanding the three major needs most victims have after a crime has been committed: the need to feel safe, the need to express their emotions, and the need to know "what comes next." (Note, the general tips provided here and throughout this guidebook are advisory only and should be considered in conjunction with your agency's own specific protocols on responding to victims of crime.)

General Tips on Responding to Victims' Three Major Needs

Victims' Need To Feel Safe

People often feel helpless, vulnerable, and frightened by the trauma of their victimization. As a first responder, you can address victims' need to feel safe by following these guidelines:

- Introduce yourself to victims by your name and title. Briefly explain your role and duties.

- Reassure victims of their safety and of your concern for them by being attentive to your own words, posture, mannerisms, and tone of voice. Although this may seem to go without saying, it can easily be forgotten in the heat or distractions of the moment. Say to victims, "You're safe now" or "I'm here now." Also, use body language to show concern, such as nodding your head, using natural eye contact, placing yourself at the victims' level rather than standing over victims who are seated, keeping an open stance rather than crossing your arms, and speaking in a calm, empathetic tone of voice.

- Ask victims to tell you in just a sentence or two what happened. Let victims know that you will conduct a full interview soon. Ask if they have any physical injuries. Take care of victims' medical needs first.

- Offer to contact a family member or friend; your agency's victim services unit, if such a unit exists; or a crisis counselor for victims.

- Be mindful of victims' privacy during your interview. Conduct the interview in a place where victims feel comfortable and secure.

- Ask simple questions that allow victims to make decisions, assert themselves, and regain control over their lives. Examples: "Would you like anything to drink?"; "May I come inside and talk with you?"; and "How would you like me to address you?"

- Ask victims about any special concerns, accommodations, or needs they may have.

- Provide a "safety net" for victims before leaving them. Make telephone calls and pull together personal and professional support for victims. Develop and give victims a pamphlet that explains "victims' rights" and lists resources available for further help or information. This pamphlet should include contact information such as your agency's victim services unit, if one exists; local crisis intervention centers and support groups; the prosecutor's and victim-witness assistance offices; the state crime victim compensation program; and other nationwide services, including toll free hotlines listed in this guidebook's Directory of National Service Providers section. Urge victims to contact and utilize these services for help.

- Give victims—in writing—your name and information on how to reach you. Encourage them to contact you if they have any questions or if you can be of further help.

Victims' Need To Express Their Emotions

Victims need to air their emotions and tell their story after the trauma of the crime. They need to have their feelings accepted and their story heard by a nonjudgmental listener. In addition to fear, victims may have feelings of self-blame, anger, shame, sadness, or denial. Their most common response is "I can't believe this happened to me." Emotional distress may surface in seemingly peculiar ways, such as laughter or an expressionless face. Sometimes victims feel rage at the sudden, unexpected, and uncontrollable threat to their safety and lives. This rage can even be directed at the people who are trying to help them—including law enforcement officers, for not arriving at the scene of the crime sooner. You can facilitate victims' need to express their emotions by following these guidelines:

- Do not interrupt or try to cut short victims' expression of their emotions.

- Observe victims' body language, such as their posture, facial expression, tone of voice, gestures, eye contact, and general appearance. This can help you understand and respond to what victims are feeling as well as to what they are saying.

- Assure victims that their emotional reactions to the crime are not uncommon. Sympathize with victims by saying "You've been through something very frightening. I'm sorry"; "What you're feeling is completely natural"; or "This was a terrible crime. I'm sorry it happened to you."

- Counter any self-blame by victims and tell them "You didn't do anything wrong. This was not your fault."

- Talk with victims as individuals. Do more than just "take a report." Sit down and place your notepad aside momentarily. Ask victims how they are feeling, and listen.

- Say to victims, "I want to hear the whole story, everything you can remember, even if you don't think it's important."

- Ask open-ended questions. Avoid questions that can be answered with a yes or no. Ask questions such as "Can you tell me what happened?" or "Is there anything else you can tell me?"

- Show that you are actively listening to victims through your facial expressions, body language, and comments such as "Take your time; I'm listening" and "We can take a break if you like; I'm in no hurry."

- Refrain from interrupting victims while they are telling their story.

- Repeat or rephrase what you think you heard victims say. Examples: "Let's see if I understood you correctly. Did you say . . . ?"; "So, as I understand it, . . . "; or "Are you saying . . . ?"

Victims' Need To Know "What Comes Next"

Victims often have concerns about their role in the investigation of the crime and in the legal proceedings. They may also be concerned about issues such as media attention on themselves and their ability to pay for medical care or property damage. Some of their anxiety may be alleviated if victims know what to expect in the aftermath of the crime. This information will also help victims prepare themselves for upcoming stressful events and disruptions in their lives related to the crime. You can respond to this need of victims to know "what comes next" by following these guidelines:

- Explain to victims what you are doing as well as the law enforcement procedures for tasks that are pending, such as the filing of your report, investigation of the crime, and the arrest and arraignment of a suspect.

- Tell victims about forthcoming law enforcement interviews or other kinds of interviews they can expect.

- Discuss the general nature of any medical forensic examinations that the victim may be asked to undergo and the importance of these examinations for law enforcement.

- Let victims know what specific information from the crime report will be available to news organizations and the likelihood of the media releasing any of this information.

- Counsel victims that lapses of concentration, memory losses, depression, and physical ailments are natural reactions for crime victims.

Encourage victims to reestablish regular routines as quickly as possible to help speed their recovery.

- Develop and give to victims a pamphlet that explains "victims' rights" and lists resources available for help and information. This pamphlet should include contact information such as your agency's victim services unit, if one exists; local crisis intervention centers and support groups; the prosecutor's and victim-witness assistance offices; the state crime victim compensation program; and other nationwide services, including toll free hotlines listed in this guidebook's Directory of National Service Providers section. Urge victims to contact and utilize these services for help.

- Advise victims as to what, if anything, they need to do next.

- Ask victims if they have any questions. Provide victims—in writing—with the incident referral number and your telephone number, and encourage them to contact you if you can be of further assistance. Follow up by providing victims with a free copy of the incident report as well as any arrest reports.

SECTION II

FIRST RESPONSE TO INDIVIDUAL TYPES OF CRIME VICTIMS

Older Victims

Background

When older people are victimized by crime, they may suffer worse physical, psychological, and financial injuries than other age groups. For example, when victims who are 65 years of age or older are injured in a violent crime, they are about twice as likely to suffer serious physical injury and to require hospitalization as any other age group.[1] Because the physiological process of aging brings with it a decreasing ability to heal after an injury, older people may also never fully recover physically or psychologically from the trauma of their victimization. In addition, this trauma may be worsened by their financial situation. Many older people live on fixed incomes and may be unable to afford the services that could help them in the aftermath of a crime.

It is understandable, therefore, why older people are often so fearful of crime. And this fear can be compounded by a number of other concerns that older people may face after a crime. They may doubt their ability to meet the expectations of law enforcement and worry that officers will think they are incompetent. They may worry that family members, upon learning of their victimization, will also think they are incompetent and belong in a nursing home. They may experience feelings of guilt for having "allowed" themselves to be victimized. They may fear retaliation by the offender, who may also be their caregiver, for having reported the crime. And older victims may be anxious about their own welfare, ashamed of their situation, and fearful of the consequences to their family member if they report abuse by that family member. Depending on your approach as a first responder, you can do much to reduce the fear, lower the anxiety, and restore the confidence of older victims with these concerns, and help them to maintain their dignity.

Finally, while some older people experience health or disability issues, many older people are healthy and active and do not have any physical or cognitive limitations that will require accommodations from you. It is important, therefore, never to assume that older people are frail or have a disability based solely on their age. Instead, it is best to ask older victims—like all victims— what special assistance, if any, they need from you as a first responder.

Tips on Responding to Older Victims

- Be attentive to whether victims are tired or not feeling well.

- Give victims time to collect their thoughts before your interview.

- Ask victims if they are having any difficulty understanding you. Be sensitive to the possibility that they may have difficulty hearing or seeing, but do not automatically assume that victims have a specific disability. Ask victims if they have any special needs, such as eyeglasses or hearing aids.

- Ask victims if they would like you to contact a family member, friend, or caregiver.

- Be alert for signs of domestic violence, elder abuse, or neglect as victims are sometimes abused by their spouse, children, relatives, or caregivers. The presence of these persons could, therefore, inhibit victims from fully describing the crime to you.

- Give victims adequate time to hear and comprehend your words during the interview.

- Ask questions one at a time, and wait for a response before proceeding to the next question. Repeat key words and phrases. Ask open-ended questions to ensure that you are being understood.

- Try to reduce or minimize the stressors and pressures on victims. Be patient. Give victims frequent breaks during your interview.

- Consider conducting a preliminary interview initially and following up the next day for more detailed information.

- Avoid subjecting victims to multiple interviews whenever you and other service providers can come together for a single interview.

- Respect the dignity of victims by including them in all decisionmaking conversations occurring in their presence.

- Provide enhanced lighting if victims need to read or write down anything. Make sure that all print in written materials is both large enough and dark enough for victims to read.

- Write down for victims—or give them printed information that explains—important points you communicate verbally so they can refer to this information later.

- Be mindful that victims may have difficulty reading or writing.

- Understand that the recollections of some older people may surface slowly; additionally, they may have memory loss or dementia. Do not pressure victims to recall events or details; rather, ask them to contact you if they remember something later.

- Reconnect with victims—as they may not initiate further contact with you or other service providers—to check on their physical and psychological condition and to obtain further information about the crime.

- Focus on the goals of restoring confidence to and maintaining the dignity of older victims in all your comments and interactions with them, their families, caregivers, and other service providers involved in the case.

- Remember, never assume that older victims have any disabilities; but, when appropriate, refer elsewhere in this guidebook for tips on responding to victims whose needs may fall under the following sections of the book: Victims Who Have Alzheimer's Disease, Victims With Blindness or Vision Impairment, Victims Who Are Deaf or Hard of Hearing, and Victims With a Disability Affecting Physical Mobility.

Child Victims

Background

Approximately one out of every four victims of crime in the United States is a child,[2] and homicide is the leading cause of non-illness related death of children under age 5.[3] Regardless of their race or social class, children are victimized at higher rates than adults in both urban and rural areas.[4] Some children are especially vulnerable to victimization, including those who are shy, lonely, and compliant; those who are labeled "bad kids"; those who are preverbal and very young; and those who have physical, emotional, or developmental disabilities.[5]

When children are victimized by crime, their psychological passage through the natural stages of growing up can be disrupted. In addition, the child—and later the adult—may have to cope with the trauma of the victimization again and again in each developmental stage of life.

Furthermore, child victims suffer not only the physical and emotional trauma of their victimization but, once the crime is reported, the trauma of being thrust into the stressful "adult" world of the criminal justice system. There, adults who were unable to protect them in the first place are responsible for restoring the child victims' sense that there are safe places where they can go and safe people to whom they can turn.

Children need to be assured that their well-being is of supreme importance to adults. As a law enforcement officer, your age-appropriate first response to children can be critical in how they initially experience their victimization and may even affect whether the crime will have a minimal or chronic impact on their life.

Tips on Responding to Child Victims

- Choose a secure, comfortable setting for interviewing child victims, such as a child advocacy center, if available, or other "child friendly"

First Response to Individual Types of Crime Victims

environment. Allow time for the child to establish trust and rapport with you. The following tips are helpful when working with child victims:

- ❑ Preschool children (ages 2 through 6) are most comfortable at home—assuming no child abuse takes place there—or in a very familiar environment. A parent or other adult the child trusts should be nearby.

- ❑ Elementary school-age children (ages 6 through 10) are sometimes reluctant to disclose information if they believe that they or their parents could "get in trouble." For this reason, the presence of a parent is usually not recommended. However, a parent or other adult the child trusts should be close by, such as in the next room.

- ❑ Preadolescents (ages 10 through 12 for girls and 12 through 14 for boys) are peer oriented and often avoid parental scrutiny. They may be more relaxed if a friend or perhaps a friend's parent is nearby.

- ❑ Adolescents (generally ages 13 through 17) are concerned about betraying their peers. It may be necessary to interview them in a setting with no peers around.

■ Be aware that children tend to regress emotionally during times of stress, acting younger than their age. For example, 8-year-olds may suck their thumb.

■ Talk in language appropriate to victims' ages and, especially with young children, do not use jargon, long sentences, or a lot of pronouns that can be confusing, like "she," "he," or "they." Remember your own childhood and try to think like the child victim, but avoid baby talk.

■ Assure preschool and elementary school-age victims that they have done nothing wrong and that they are not in trouble; young children often are afraid that they will be blamed for problems and may have been told by the offender that they would be blamed.

■ Realize that children are more likely than adults to blame themselves for abuse, particularly if the offender is someone with whom the child has a close relationship. Also, know that most children do not make up stories of child abuse, and false allegations are the exception. It is far more likely that child victims will lie to conceal abuse and protect the offender.

- Be consistent in the terms and language you use and repeat important information often.

- Ask open-ended questions to make sure victims understand you.

- Maintain a nonjudgmental attitude and patiently empathize with victims. Children need to communicate what happened and to have the reality of their experiences validated.

- Compliment victims frequently on their good behavior and for answering your questions, as well as for telling you when they do not understand a question. Elementary school-age children are especially affected by praise. Be careful, however, not to praise child victims on the substance of their answers. At the prosecution of a criminal defendant, such praise could be deemed suggestive questioning by you of the child.

- Be mindful of the limited attention span of children and their tendency to disclose facts regarding traumatic events over time. Observe child victims for signs that they are tired, restless, or cranky. When interviewing preschool children, consider doing a series of short interviews rather than a single, lengthy one. Also, consider postponing the interview until the victim has had a night's rest. However, do not wait too long before interviewing preschool children; victims at this age can have difficulty separating the details of their victimization from later experiences.

- Appreciate that children, like adults, find it upsetting to talk about traumatic events. Young children particularly may "relive" their victimization and feel the associated emotions again, thereby intensifying their trauma. Thus, minimize the number of times that victims must be interviewed.

- Include victims, whenever possible, in decisionmaking and problem-solving discussions. Identify and patiently answer all their questions. Reduce victims' anxiety by explaining the purpose of your interview and by preparing them, especially elementary school-age children, for what will happen next.

- Show sensitivity in addressing sexual matters with preadolescent and adolescent children. Because they are developing their sexual identity, their self-consciousness and a limited vocabulary can make such conversations embarrassing for them. Conversely, do not assume that victims, especially elementary school-age children, are as knowledgeable about sexual matters as their language or apparent sophistication might indicate.

- Know that teenagers are the age group least likely to report their violent victimization,[6] and that they are the most likely to be victimized.[7] Victimization can intensify the normal adolescent insecurities of being different from or not as "tough" as one's peers. Respect victims and empathize with their concerns so that these negative feelings do not lead them to despair or to seek revenge. Also, consider providing a referral to professional counseling.

- Have compassion for victims. Children's natural abilities to cope are aided immensely by caring adults.

- Do not neglect to comfort the nonoffending parent. Although the immediate victim is the child, parents of the victim should be referred to agencies that can assist them in coping and that can advise them on what to expect and how to talk with their child about the victimization.

Victims Who Have a Disability
Background

Anyone can be victimized by crime but people who have a disability are more vulnerable to crime than others in society.[8] People with a mental disability can be less able to recognize and avoid danger, and people with a physical disability can be less able to protect themselves or escape harm. Furthermore, victims of crime who have a disability can be less able to contact law enforcement and—without accommodations for their disability—assist in the investigation of the crime.

About one in five people in the United States has a mental or physical disability,[9] and for almost half of this population the disability is severe.[10] These disabilities come in many forms but they all affect either a person's mental functioning, such as the ability to reason and exercise good judgment, or a person's sensory or physical abilities, such as the ability to see, hear, and walk.

One reason that the risk of criminal victimization for people with a disability is much greater than for those without disabilities is that perpetrators specifically target this population under the assumption that victims will be unable to escape or report the crime. In addition, people who have a disability are often victimized repeatedly by the same perpetrators, and these perpetrators may include their caregivers.

Although most issues confronted by crime victims who have a disability are concerns that affect crime victims in general, there are still important differences in how to approach and help victims with a disability. The information presented in this section provides an overview of some of these differences and briefly illustrates how to better serve crime victims who have a disability.

General Tips on Responding to Victims Who Have a Disability

A lack of personal familiarity with individuals who have a disability may naturally cause you to feel self-conscious and uncertain in your response to victims of crime with disabilities. On the other hand, a person's disability may not be obvious, so watch victims carefully for signs of any disability. Do not hesitate to ask victims if they have any special needs. Yet, also be sure to appreciate that people with disabilities are not a homogeneous group, but individuals with differing capabilities and needs. As a first responder, you can promote effective communication and thereby better serve victims of crime who have a disability by observing the following guidelines:

- Reflect on some of the stereotypes that exist about people who have a disability. Negative attitudes may be the greatest impairment for people with disabilities.

- Avoid labeling or defining victims by their disability. Instead, use "people-first" language that emphasizes the person, not a disability. For example, referring to the victim as "a disabled woman" implies that she is primarily disabled and secondarily a woman. Referring to her as "a woman with a disability" would be preferred because this phrase portrays an image of a female adult victim who happens to have a disability. Similarly, saying "the man has schizophrenia" is preferable to "the man is a schizophrenic." In other words, the victim *has* a disabling condition, not *is* that condition.

- Use the word "disability" rather than "handicap." A disabling condition need not be handicapping. People who use a wheelchair, for instance, have a disability, but they are not handicapped by stairs when a ramp is available.

- Ask victims directly how the two of you can most effectively communicate with each other, how they wish their disability to be characterized, and how you can best assist them. Most victims would prefer to answer these few questions upfront rather than endure your uneasiness or be uncomfortable themselves throughout an entire interview. Your respectful and sensitive questions will ensure that the language you use and the accommodations you make are appropriate, not detrimental.

- Relax, and do not be embarrassed when you use common expressions that seem related to a victim's disability, such as saying "Do you see my point?" to a person with a vision impairment; "I'm waiting to hear back from her," to a person with a hearing loss; or "I need to run over there," to someone who uses a wheelchair. Victims know what you mean and should not take offense.

- Recognize that the presence of someone familiar to victims or a person knowledgeable about their disability may be extremely important for victims and helpful during your interview. But remember that family members, personal care attendants, and service providers could themselves be the offenders or be protecting the offenders. Their presence, therefore, may inhibit victims, out of fear of retaliation, from fully describing the crime to you.

- Do not act on your curiosity about victims' disabilities. Restrict your questions to those necessary to accommodate victims' needs; focus on the issues at hand, not the disability.

- Avoid common expressions of pity such as "suffering from" Alzheimer's disease or "a victim of" mental illness.

- Speak directly to victims, even when they are accompanied by another person. People with disabilities are sometimes assumed to be incapable of making decisions for themselves, and you do not want to give the impression that you may think this way.

- Listen to your tone of voice and monitor your behavior to make sure that you are not talking down to victims, coming across in a condescending manner, or treating victims as children.

- Do not express admiration for the abilities or accomplishments of victims in light of their disability.

- Be mindful of the underlying painful message communicated to victims by comments such as "I can't believe they did this to someone like you"; "She's disabled and he raped her anyway"; or "To steal from a blind man, that's got to be the lowest." Such phrases can send the wrong message—that you consider people who have disabilities as "less than" complete human beings.

- Document victims' disabilities in your incident report, as well as their individualized communication, transportation, medication, and other accommodation needs.

- Make sure that victims are in a safe environment before you leave the scene. Again, recognize that victims' family members, personal care attendants, and service providers could themselves be the offenders, and that victims may need an alternate caregiver or shelter. Contact a victim advocate whenever possible for immediate victim services and followup.

- Never assume that people with disabilities suffer less emotional trauma and psychological injury than other crime victims.

- Familiarize yourself with your state and municipal statutes that address crimes against persons with disabilities.

- Be aware that federal law requires—with few exceptions—that law enforcement make reasonable modifications to policies, practices, and procedures, where needed to accommodate crime victims who have a disability, unless doing so would fundamentally alter the service, program, or activity the agency provides. (See the section on the Americans with Disabilities Act of 1990 and Section 504 of the Rehabilitation Act of 1973 in this guidebook for more information on the federal law and responding to victims who have a disability.)

Americans with Disabilities Act of 1990 and Section 504 of the Rehabilitation Act of 1973

Two federal laws—the Americans with Disabilities Act of 1990 (ADA) and Section 504 of the Rehabilitation Act of 1973—prohibit discrimination on the basis of a disability. Title II of the ADA applies to state and local government entities. Section 504 applies to recipients of federal financial assistance, including recipients of grants from the U.S. Department of Justice (DOJ).

An individual with a disability is defined by the ADA and Section 504 as a person who (1) has a physical or mental impairment that substantially limits one or more major life activities, (2) has a record of such an impairment, or (3) is regarded as having such an impairment.

Both Title II of the ADA and Section 504 require—with few exceptions—that first response officers provide victims of crime who have a disability with an equal opportunity to benefit from and participate in all programs, services, and activities of the law enforcement agency. In addition, officers must provide for equally effective communication with victims who have a disability. Law enforcement, therefore, is required to make reasonable modifications to policies, practices, and procedures where needed to accommodate crime victims who have a disability, unless doing so would fundamentally alter the service, program, or activity the agency provides.

For more information about your responsibilities under the ADA and Section 504, call DOJ's ADA Information Line at 1–800–514–0301 or DOJ's Office of Justice Programs, Office for Civil Rights at 202–307–0690.

Victims Who Have Alzheimer's Disease

Background

Alzheimer's disease is a brain disorder that breaks down the connections between nerve cells in the brain. Among older people, Alzheimer's is the most common form of dementia,[11] a progressive, irreversible condition that is characterized by a loss of mental and cognitive abilities as well as changes in personality and behavior. Outward signs of Alzheimer's disease may not be apparent in a person until the disease reaches its advanced stages. Initially, Alzheimer's causes people to forget recent events and familiar tasks. Gradually, the disease destroys a person's memory and ability to reason, think abstractly, use language to communicate, and perform daily activities. Alzheimer's may also cause mood disturbances, including anxiety, suspiciousness, agitation, delusions, and hallucinations. Eventually, people with the disease are no longer able to care for themselves.

First responders may observe the following common symptoms in people with Alzheimer's disease:

- Use of nonsensical words in speaking.

- Disoriented sense of time and place.

- Poor judgment. For example, wearing an overcoat in the summer or a nightgown to go shopping.

- Wandering or becoming lost and not knowing where one lives.

- Rapid mood swings, due to anxiety, suspiciousness, or agitation.

- Blank facial expression.

- Walking gait characterized by slow, sliding movements without lifting the feet.

Be aware that Alzheimer's disease can occur as early in age as a person's 30s and 40s. However, most of the estimated 5.1 million Americans with Alzheimer's[12] are age 65 and older. Thirteen percent of persons age 65 and over,[13] and

First Response to Individual Types of Crime Victims

nearly half of persons over 85,[14] have the disease. Seventy percent of people with Alzheimer's and other dementias live at home,[15] not in a caregiving facility; and 60 percent of people with Alzheimer's will eventually wander[16] and may become lost. Thus, with the rapid increase in the number of older persons in the U.S. population that will begin in 2011—when the first of the baby boom generation turns 65—and continue for many years thereafter,[17] the frequency with which first responders will encounter victims who have this disability will continue to increase.

Tips on Responding to Victims Who Have Alzheimer's Disease

- Approach victims from the front and establish and maintain eye contact (when you know in advance that the victim has Alzheimer's disease). Introduce yourself as a law enforcement officer and explain that you have come to help. Due to their impaired short-term memory, victims may repeatedly ask who you are. Be prepared to patiently reintroduce yourself several times.

- Request to see identification if you suspect that victims have Alzheimer's disease. In addition, notice if victims have a Safe Return® bracelet, necklace, lapel pin, key chain, or label inside their clothing collar. Safe Return identification provides the first name of a person bearing this ID, indicates that he or she has a memory impairment, and gives the 24-hour, toll free number for the Alzheimer's Association Safe Return program. This program is a nationwide participant registry that contains the full name of the registrant, a photograph, identifying characteristics, medical information, and emergency contact information. When you call the program's incident line at 1–800–572–1122, a Safe Return clinician will contact the registrant's caregivers.

- Keep in mind that persons with Alzheimer's disease who have wandered are at high risk of dehydration and hypothermia. Also, many people with Alzheimer's have serious medical conditions and are on medications that they probably will not have with them. Provide for transport to a hospital emergency room whenever medical attention is indicated.

- Treat victims with dignity. The deterioration of their mental abilities does not mean that victims are without feelings.

First Response to Victims of Crime

- Move victims away from crowds and other noisy areas. These environments can cause restlessness, pacing, agitation, and panic in people who have Alzheimer's. Also, turn off your car's flashing lights and lower the volume on your radio.

- Establish a one-on-one dialogue with victims. Talk in a low-pitched, reassuring tone, looking into the victim's eyes. Alzheimer's shortens one's attention span and increases mistrust. Your calm support can make victims less agitated, less suspicious of you, and less panicky. Speak slowly and clearly, using short, simple sentences and familiar words. Repeat your statements. Accompany your words with gestures when this can aid in communication, but avoid sudden movements.

- Include victims in all conversations, out of respect and so you will not arouse their mistrust and suspicion of your intentions.

- Explain your intended actions prior to beginning them. If victims are agitated or panicky, gently pat them or hold their hand, but avoid physical contact that could seem restraining.

- Anticipate difficulties in making yourself understood. Do not assume that victims understand you or are capable of answering your questions and complying with your instructions.

- Give simple, step-by-step instructions and, whenever possible, a single instruction. For example, "Please sit here. I'll take care of everything." Avoid multiple, complex, or wordy instructions such as "Please sit here. Don't get up or go anywhere. I'll take care of everything. Just wait for me to come back." Also, try substituting nonverbal communication for verbal instructions. For example, if you want victims to sit down, show them by sitting down yourself.

- Ask one question at a time. Yes or no questions are better than questions that require victims to recall and recite a sequence of events. Be prepared for answers that are confusing and that keep changing. If victims' words are unintelligible, ask them to point, gesture, or otherwise physically communicate their answers.

- Never challenge victims' logic or reasoning.

- Do not leave victims alone; they may wander away.

- Encourage victims' family and other caregivers to contact the Safe Return program's nonemergency number at 1–888–572–8566 to register victims if they are not already listed in the program's registry. Advise these caregivers that wandering is a life-threatening behavior.

- Find emergency shelter for victims with the help of your area's Agency on Aging or other local social service agencies, including a local chapter of the Alzheimer's Association, if no other caregivers can be found. Look in the telephone book for these service providers or, to locate an Alzheimer's Association chapter, call the national office at 1–800–272–3900.

Victims Who Have a Mental Illness
Background

Mental illness encompasses a number of distinct brain disorders—such as manic-depressive illness, schizophrenia, major depression, and severe anxiety—that disrupt a person's mood balance, thought processes, memory, sensory input, feelings, and ability to reason and relate to others. An estimated 6 percent of Americans 18 years of age and older, or 1 in 17, have a "serious" mental illness[18] that diminishes their capacity to meet the ordinary demands of life. Among children between the ages of 9 and 17, approximately 5 to 9 percent have a "serious emotional disturbance."[19]

Mental illness should not be confused with mental retardation. People with a mental illness are usually of normal intelligence but may have difficulty functioning at normal levels due to their illness.

The symptoms of mental illness vary from individual to individual depending on the type and severity of the disorder. Many symptoms are not readily observable from outward appearances but are noticeable in conversation. Although a first responder cannot be expected to recognize specific types of mental illness, the following symptoms are indicative that a victim may have a mental illness:

- Accelerated speaking or hyperactivity.

- Delusions and paranoia. For example, victims may believe, falsely, that they are a famous person or that others are trying to harm them.

- Hallucinations, such as hearing voices or seeing, feeling, or smelling imaginary things.

- Depression.

- Inappropriate emotional response. For example, silliness or laughter at a serious moment.

- Unintelligible conversation.

- Loss of memory. Not ordinary forgetfulness, but rather an inability to remember the day, year, or where the person is.

- Catatonia, which is characterized by a marked lack of movement, activity, or expression.

- Unfounded anxiety, panic, or fright.

- Confusion.

Anyone who is a victim of crime may be traumatized and experience the victimization as a crisis. But for people with a mental illness, this crisis may be experienced more profoundly. The following guidelines can assist law enforcement in better responding to crime victims who have a mental illness.

Tips on Responding to Victims Who Have a Mental Illness

- Approach victims in a calm, nonthreatening, and reassuring manner. Victims may be overwhelmed by delusions, paranoia, or hallucinations. They may be afraid of you or feel threatened by you. Introduce yourself

First Response to Individual Types of Crime Victims

personally by your name, first, then your rank and agency. Make victims feel that they are in control of the situation.

- Determine whether victims have a family member, guardian, or mental health service provider who helps them with daily living. If they do, contact that person immediately. But remember that these persons could themselves be the offenders, or may try to protect the offenders.

- Contact the local mental health crisis center immediately if victims are extremely agitated, distracted, uncommunicative, or displaying inappropriate emotional responses. Victims may be experiencing a psychiatric crisis.

- Ask victims if they are taking any medications and, if so, the types prescribed. If they are unable to provide this information, ask their family member, guardian, or mental health service provider. Make sure that victims have access to water, food, and toilet facilities as side effects of the medications can include thirst, urinary frequency, nausea, constipation, and diarrhea.

- Conduct your interview in a setting that is free of people or distractions upsetting to victims. If possible, only one officer should interview victims.

- Keep your interview simple and brief. Be friendly and patient and offer encouragement when speaking to victims. Understand that a logical discussion may not be possible on some or all topics.

- Remember that even victims who are experiencing delusions, paranoia, or hallucinations may still be able to accurately provide information that is outside of their false system of thoughts, including details related to their victimization, as well as informed consent to medical treatment and forensic exams.

- Avoid the following conduct in your actions and behavior with victims:

    ❑ Circling, surrounding, closing in on, or standing too close to victims.

    ❑ Concealing your hands.

- ❏ Sudden movements or rapid instructions and questioning.

- ❏ Whispering, joking, or laughing.

- ❏ Direct, continuous eye contact; forced conversation; or signs of impatience.

- ❏ Any touching.

- ❏ Challenges to, or agreement with, victims' delusions, paranoia, or hallucinations.

- ❏ Inappropriate language, such as "crazy," "psycho," and "nuts."

■ Back off and allow victims time to calm down before intervening if they are acting excitedly or dangerously but there is no immediate threat to anyone's safety. Outbursts are usually of short duration.

■ Break the speech pattern of victims who talk nonstop by interrupting them with simple questions, such as their birth date or full name, to bring compulsive talking under control.

■ Do not assume that victims who are unresponsive to your statements cannot hear you. Do not ignore them or act as if they are not present. Be sensitive to all types of response, including victims' body language.

■ Acknowledge victims' paranoia and delusions by empathizing with their feelings; but neither agree nor agitate victims by disagreeing with their paranoid or delusional statements. For example, if victims tell you that someone wants to hurt them, reply with "I can see that you're afraid. What can I do to make you feel safer?" Be mindful, however, that victims who say that others are trying to harm them may indeed be the victims of stalking or other crimes.

■ Understand that hallucinations are frighteningly real to victims, and never try to convince victims that their hallucinations do not exist. Instead, reassure victims that the hallucinations will not harm them and may disappear as their stress lessens.

First Response to Individual Types of Crime Victims

- Assess victims' emotional state continuously for any indications that they may be a danger to themselves or others.

- Be honest with victims. Getting caught by victims in your well-intentioned untruth will only increase their fear and suspicion of you.

- Arrange for victims' care by a family member, guardian, or mental health service provider before leaving them. But, again, remember that these persons could themselves be the offenders, or may try to protect the offenders.

Victims With Mental Retardation
Background

Mental retardation is a disability affecting the brain and its ability to process information. People with mental retardation have difficulty learning and are below average in intelligence. They have problems with judgment and in their abilities to focus, understand, and reason.

Between 1 and 3 percent of Americans have mental retardation.[20] But persons with mental retardation appear to represent much more than 3 percent of crime victims; they also appear to be at higher risk for victimization than people without disabilities.[21]

Most people with mental retardation are only mildly affected[22] and look no different from anyone else, making mental retardation difficult for first responders to recognize.

Furthermore, people with mental retardation may try to hide their disability or pretend to have greater capabilities than they actually possess. There are, however, questions you can ask and traits you can watch for when attempting to determine if a crime victim has mental retardation:

- Ask victims where they live, where they work or go to school, and if they have someone who helps them when they have a problem. Victims' answers will let you know if they live with their parents or in a group home, are employed in a vocational rehabilitation setting or attend special education classes, and if they have a social worker.

- Ask victims for directions to their home and to read or write something for you. Observe victims for any signs that they are having difficulty understanding you; listen to whether they have a limited vocabulary; and watch for any frustrations they may exhibit as they respond to these simple requests.

Ultimately, there is often no way for the first responder to know whether a crime victim has mental retardation. Persons with the disability can vary widely in their capabilities and skills. If you suspect this disability, proceed as though the victim does have mental retardation. In so doing, you can best ensure effective communication and an optimum response to the victim's needs.

Tips on Responding to Victims With Mental Retardation

- Show the same level of respect to crime victims with mental retardation that you show to all victims.

- Introduce yourself, first, as a law enforcement officer, followed by your agency and name. People with mental retardation have usually been taught that officers are their friends and can be trusted to keep them safe. (However, victims of ongoing abuse are sometimes told lies by their abusers about law enforcement and other service providers.)

- Avoid mentioning victims' disabilities in front of them. If this is not possible, refer to the victim as a person with mental retardation or a cognitive disability. Never use the word "retarded."

- Do not assume that victims are incapable of understanding or communicating with you. Most people with mental retardation live independently or semi-independently in the community, so a fairly normal conversation should be possible.

- Create a safe atmosphere, limit distractions, and establish a trusting rapport with victims before interviewing them.

First Response to Individual Types of Crime Victims

- Be mindful of whether or not victims are "competent" to give or withhold consent to medical treatment and forensic examinations, notification of next of kin, and other services; but do not assume that victims are incompetent.

- Explain written information to victims and offer to help them fill out paperwork.

- Ask victims if there is anyone whom they would like you to call to be with them during your interview. But remember that family members, service providers, and others can have a vested interest in the interview. They could themselves be the offenders, or may try to protect the offenders.

- Allow adequate time for your interview and give victims a break every 15 minutes.

- Treat adult victims as adults, not as children.

- Speak directly and slowly to victims, keeping your sentences short and words simple. Listen to how victims talk and match your speech to their vocabulary, tempo, and sentence structure.

- Separate complex information into smaller parts and use gestures and other visual props to make yourself understood. Do not overload victims with too much information.

- Recognize that victims may be eager to please you or be easily influenced by you. They may say what they think you want to hear; so be careful not to ask leading questions.

- Use open-ended questions or statements that cannot be answered with a yes or no, such as "Tell me what happened." Let victims "lead the interview" as they disclose information.

- Help victims understand your questions by giving them points of reference. For example, ask "What color was the man's hair?" rather than "What did the man look like?" and "Did the fight start before or after lunch?" instead of "When did the fight start?"

- Give victims at least 30 seconds to respond to an instruction or question. If they do not respond, or if they respond inappropriately, patiently repeat yourself, using different words. Also, have victims state in their own words what they understood you to say.

- Repeat the last phrase of victims' statements in the form of a question to help victims stay focused during your interview and to transition them through a sequence of events. For example, ask "He hit you?"; "You fell down?"; and "Then you tried to run away?"

- Avoid questions that can confuse victims or that require a great deal of mental reasoning or insight. Examples of types of questions to avoid include "Why do you think she did this to you?"; "Do you have any idea of what was really going on?"; or "What made you do that?"

- Know that resources exist to help you respond to crime victims with mental retardation. Look in the telephone book under "social service organizations"; contact your local United Way or local chapter of The Arc (an organization for people with mental retardation); or call the national office of The Arc at 1-800-433-5255 for assistance on how best to serve victims with mental retardation.

Victims With Blindness or Vision Impairment
Background

The ability to see exists along a wide continuum from sighted to partially sighted to blind. Complete blindness, "legal" blindness, and vision impairment affect an estimated 14 million people in the United States ages 12 and older.[23] Complete blindness is rare,[24] and means totally without sight. Legal blindness denotes the condition whereby a person is unable to see at 20 feet what someone with normal vision can see at 200 feet. And people with vision impairment may have partial sight that allows them to get around without much difficulty. However, they need adaptive methods to read and write because of the impairment.

Blindness—whether complete or legal—and vision impairment become more common with advancing age. More than two-thirds of people with this disability are older than 65.[25] Thus, it is predicted that the number of persons with the disability will increase substantially over the next few decades as baby boomers enter their older years. Whatever a crime victim's age, however, first responders can effectively meet the needs of victims with blindness or vision impairment by following the guidelines listed below.

Tips on Responding to Victims With Blindness or Vision Impairment

- Introduce yourself immediately as a law enforcement officer when you approach victims and have others who are present introduce themselves, including children. These introductions let the victim know who is present and where they are situated, and also help the victim recognize voices during your interview. In addition, mention the presence of a dog, cat, or other pet to protect victims from tripping over the animals or being startled by them.

- Tell victims your name, badge number, and the telephone number of your dispatcher when responding to victims who are alone, and support them in verifying your identity.

- Describe the chair and seating arrangements when assisting victims in sitting down; and place their hand (after obtaining permission) on the back or arm of the chair. No further assistance is necessary.

- Do not speak loudly. Most people with blindness or vision impairment are not hard of hearing.

- Identify the person to whom you are speaking when talking with a group of people. Exactly which person you are speaking to may not be apparent to victims who are blind or who have a vision impairment.

- Let victims know when you or other persons step away (and return) during a conversation.

- Avoid lapses of conversation in your interview without first informing victims of why you need to be silent; for example, tell victims that you are writing. Also, project attentiveness, concern, and compassion through your voice and choice of words. Remember that victims cannot see your facial expressions or body language to ascertain whether you are listening to them and interested in what they have to say.

- Offer to fill out forms and read aloud written information for victims. Explain what printed materials you are providing and, upon request, make the materials available in an alternative format, including large print, audiotape, computer diskette, and braille. This provision of materials in an alternative format is legally required, with few exceptions, by the Americans with Disabilities Act and Section 504 of the Rehabilitation Act of 1973.

- Never separate victims from their guide dogs or pet the dogs without permission. There is a special relationship between people who are blind and their dogs, and the dogs are working animals that must not be distracted.

- Offer your arm, instead of holding the arm of victims, if they want you to guide them in moving about. Let victims take your arm from behind, just above the elbow. In this position, they can follow the motion of your body. Walk in a relaxed manner and remember that victims will follow at a half-step behind you so they can anticipate curbs and steps.

- Orient victims to their surroundings and give cues as to what lies ahead when guiding them. Close doors to cabinets, rooms, and cars that obstruct their path. Warn of hazardous objects around them. Before going up or down stairs, come to a complete stop and inform victims about the direction of the stairs, the approximate number of steps, and the location of the handrail. Finally, make your warnings and directions specific, such as "in front of you" and "to your left," rather than giving vague references like "at the front of the room" or "beside you."

Victims Who Are Deaf or Hard of Hearing
Background

The term "deaf" is used in reference to people who are unable to hear or understand oral communications even with the aid of an amplification device.

"Hard of hearing" refers to a hearing loss severe enough to necessitate the use of such a device to effectively hear oral communications.

About 10 percent of the population in the United States has some degree of hearing loss.[26] Among older persons, one in three people over age 60[27] and half of those over 85 have a hearing loss.[28]

Whether deaf or hard of hearing, victims of crime with this disability are fully capable of cooperating with first responders. To effectively meet victims' needs, however, the officer must quickly determine the method by which victims want to communicate, and then immediately begin using it.

Tips on Responding to Victims Who Are Deaf or Hard of Hearing

- Signal your presence to victims by waving your hand or gently—so as not to startle—touching victims on the arm or shoulder if they do not notice you.

- Determine how victims want to communicate by initially communicating through writing in situations where victims do not speech- or lip-read and a sign language interpreter is not immediately available.

- Honor victims' request for a sign language interpreter. The Americans with Disabilities Act and Section 504 of the Rehabilitation Act of 1973 require that the expressed request of victims for an interpreter be given primary consideration in determining which communication aid to provide. You must always meet a victim's request unless your agency can demonstrate that another effective method of communication exists. Although the ADA does not require that interpreters be certified, they must be qualified. The national Registry of Interpreters for the Deaf, at 703-838-0030, has chapters in all 50 states that can help you locate a qualified interpreter.

- Realize that victims may not be literate in written English but may know American Sign Language (ASL).

- Never use a child to assist in communicating with victims except in emergency situations or for short-term assistance in helping you to locate an appropriate adult.

- Do not assume that because victims are wearing hearing aids that they can hear or understand you. The degree and type of hearing loss a person may have could render hearing aids of limited value, in terms of assistance with tones of speech.

- Remember in all your interactions with victims that people who are deaf or hard of hearing are visually oriented.

- Avoid shouting or speaking very slowly to make yourself understood as this distorts your speech, lip movements, and facial expressions, and may make you appear upset.

- Be mindful that not all people who are deaf or hard of hearing can speech- or lip-read, and that even for the best speech readers under the most favorable conditions, much of spoken English cannot be distinguished on the lips.

- Use gestures, mime, and props to better communicate. For example, motion toward a chair to offer victims a seat; mimic drinking from a glass to ask victims if they are thirsty; and touch your clothing, face, or hair when interviewing victims for a description of the offender.

- Do not assume that victims are unable to speak or use their voices. Never use the words "deaf mute" or "deaf and dumb." Deaf people have the ability to use their voice but may prefer not to speak because of the quality of their speech.

- Observe victims' facial expressions and other physical mannerisms very closely; much information can be gleaned by paying attention to victims' body language.

- Include victims in all conversations and describe any commotion in the area. If you look away from victims because you overhear another conversation, if you are distracted by a noise or other disturbance, or if

First Response to Individual Types of Crime Victims

you turn away from victims to talk with someone else, explain to victims exactly what you are doing or what is happening.

- When interviewing victims who are hard of hearing—or who are deaf and want to communicate by speech- or lip-reading—select a location free of distractions, interference, and, for those victims who are hard of hearing, any background noise, and follow the tips listed below:

  ❑ Face victims so that your eyes and mouth are clearly visible. Be careful not to block your mouth with your hands, or to speak while you are looking away from victims or down at your notes. Do not chew gum.

  ❑ Stand or sit between 3 and 6 feet from victims in a well-lit and glare- and shadow-free area. Refrain from all unnecessary gesturing and body movements as it is difficult for victims to speech- or lip-read if you are not physically still.

  ❑ Begin speaking only after you have the victim's attention and have established eye contact.

  ❑ Make your questions, instructions, answers, and comments short and simple.

  ❑ Speak clearly, distinctly, and slightly slower than usual but not unnaturally slow; do not exaggerate your pronunciation of words.

  ❑ Talk slightly louder than usual but never shout. Hearing aids do not transmit extremely loud tones as well as they do normal tones, and shouting distorts your lip movements.

  ❑ Be prepared to repeat yourself many times, but use different words in your reiterations. Sometimes a particular group of lip movements is difficult to read or the victim may have missed only a word or two initially and your restatement will clarify what was missed.

  ❑ Use open-ended questions and statements that require victims to answer with more than a yes or no to prevent misunderstandings. For example, say "Describe the offender to me" rather than asking "Is the offender someone you know?"

- Do not rely on victims' family members—who have probably been emotionally affected by the crime as well—or any other person with a potential bias to provide sign language interpreting for your interview.

- When communicating through an interpreter, remember that the interpreter is present solely to transmit information back and forth between the victim and you, not to explain information or give opinions. When using an interpreter, you should—

  - Stand or sit across from the victim, in a glare- and shadow-free area, with the interpreter beside you so that the victim can easily shift his or her gaze between you and the interpreter.

  - Speak at a normal volume and pace, and directly to the victim, not the interpreter. Never ask "How is he feeling?" or say "Ask her how she is feeling." Address your questions to the victim: "How are you feeling?"

  - Ask the victim, not the interpreter, to repeat or clarify an answer if you do not understand it.

  - Take breaks. Interpreting (signing) and receiving information visually can be tiring for both the victim and interpreter.

- Be aware that a "Deaf culture" exists. This culture has a language—American Sign Language (ASL)—and experiences, practices, and beliefs about itself and its connection to the larger hearing society. Victims who identify with the Deaf culture may live more isolated from and be less comfortable with the hearing society, including you, than victims for whom hearing loss is merely a physical condition.

First Response to Individual Types of Crime Victims

Victims With a Disability Affecting Physical Mobility

Background

Approximately 2.2 million people in the United States use a wheelchair,[29] and another 6.4 million use some other ambulatory aid, such as a cane, walker, or crutches,[30] according to U.S. Census Bureau data of persons 15 years old and older who are not living in an institution. The number of people with a disability affecting their physical mobility is, of course, even higher when persons under age 15 and those living in an institution are included.

People with mobility disabilities are usually quite independent and fully capable of cooperating with law enforcement. Crime victims with these disabilities do, however, often require accommodations. Therefore, to effectively serve this population, first responders should quickly determine what accommodations victims may need and provide them.

Tips on Responding to Victims With a Disability Affecting Physical Mobility

- Ask victims "May I assist you?" rather than trying to help without asking. If victims need assistance, listen to their instructions. Victims know best how you can be helpful and what is safest for them.

- Be careful about making assumptions that victims cannot do something. Instead, ask how they usually do it.

- Treat assistive devices (wheelchairs, canes, walkers, crutches) as victims' "personal space" and valuable property. Do not lean on or rest your feet against a victim's wheelchair, and do not touch any assistive device without first obtaining permission from the victim.

- Do not fall into an "up here/down there" mindset toward victims who use a wheelchair, especially when you are talking with other persons who are standing. Position yourself in front of victims, at eye level, but do not kneel. If you cannot sit down, stand far enough back from victims so that you are not towering over them and so they are not straining their neck to see you.

- Move obstacles and open/close doors so that hallways are free of barriers and so that victims can easily and safely go forward, backward, and turn around.

- Offer to assist victims if they seem to be having trouble maneuvering in tight spaces.

- Ask victims who are using a wheelchair, or who are on a wheeled cot (gurney), which direction they prefer to face when you are assisting them in going up or down stairs or steep inclines. Also, use caution not to damage the wheelchair.

- Be considerate of the extra time victims may need to move about and let them set your walking pace.

- Ask victims if they have a seating preference in situations where you are interviewing other persons at the same time. Recognize that victims may not need a special place upfront and probably do not want a segregated seating placement that draws attention and indicates that they are "different."

- Do not assume that victims who use a wheelchair can only sit in that type of chair. When other seating arrangements are available, ask victims if they would like to transfer to another chair. Always make sure that the wheelchair is locked before helping victims to transfer.

- Realize that victims can fully describe the crime, even if they are unable to physically demonstrate it.

- Never refer to victims as "crippled," "confined" to a wheelchair, "wheelchair-bound," or "handicapped."

- Ask victims if their wheelchair was damaged during the crime and, if so, assist them in getting immediate repairs or a loaner chair.

- Confirm with service providers that accommodations are available for victims before making referrals. Do not assume that providers know what architectural, transportation, and other accommodations are needed. Ask victims about their specific needs and then tell providers of accommodation requirements.

- Assure victims that you understand their need for accommodations and specifically inform them of the accommodations that will be provided.

- Be mindful that victims' personal attendant could be the offender or could be protecting the offender. Furthermore, be prepared to assist victims in locating a replacement for the attendant.

- Offer to assist victims in filling out paperwork.

- Present all informational materials to victims directly. Ask victims if they would prefer that you give the materials to another person; or if victims have a purse, backpack, or other carrying case, ask if they would like you to place the documents there.

Immigrant Victims

Background

As of 2003, the foreign-born population in the United States represented almost 12 percent of the civilian noninstitutionalized U.S. population.[31] This foreign-born population, particularly recent immigrants, can be especially vulnerable to crime. Moreover, immigrants are less likely to report their victimization to law enforcement, or access support services in the aftermath of a crime, than are members of the native-born U.S. population.[32]

In America's increasingly multicultural society, situations involving cross-cultural interaction are becoming ever more commonplace for law enforcement. But still, most officers have had only minimal experience interacting with people from other countries. Differences in cultural norms and in verbal and nonverbal modes of expression can often result, therefore, in miscommunication between first responders and immigrant victims of crime.

As new arrivals to this country, recent immigrants also face many issues—sometimes unrecognized by law enforcement—that go well beyond the more obvious cultural and language challenges. For example, based on negative experiences in their native countries, immigrants may fear or distrust law enforcement and be reluctant to call for help or cooperate with first responders trying to assist them. In addition, they may worry about their immigration status, or that of family members, and not report crimes out of concern that an investigating officer will turn them over to immigration authorities and they will be deported.

Sensitivity by officers to the complexities in communication that can result from language and cultural differences between people; empathy for the unique concerns of immigrants; and a respectful attitude toward the foreign-born and their native countries, are critical, therefore, to fostering law enforcement's rapport with immigrant victims of crime and facilitating an effective first response.

Tips on Responding to Immigrant Victims

- Be knowledgeable about Title VI of the Civil Rights Act of 1964, which prohibits recipients of federal financial assistance, including law enforcement agencies, from discriminating against people based on national origin. Discrimination occurs when, among other things, an agency fails to provide meaningful access to its programs and activities for people who have limited English proficiency (LEP). Written materials, for example, that are routinely provided in English to victims must also be translated into the languages of regularly encountered LEP victims eligible to be served or likely to be affected by an agency's programs or activities.

- Rethink any stereotypes—negative or positive—that you may have about people from different cultural backgrounds; be aware that these biases tend to reveal themselves when one is under stress or patience is running thin.

- Appreciate that every culture has its own norms as to what are appropriate communication mannerisms, such as eye contact, physical gestures, touching, and the display of emotions; do not expect victims' communication style to conform to your norms.

First Response to Individual Types of Crime Victims

- Explain your role and duties as a law enforcement officer to victims, as well as any law enforcement and court procedures relevant to their case, including the possibility of pretrial release for the offender.

- Familiarize yourself with the many varying perspectives that different cultures have regarding male/female and youth/elderly role expectations, domestic and sexual violence, public versus private or family matters, religion, and "saving face" for one's self or family. Tailor your approach to maximize your effectiveness within victims' cultural perspectives.

- Admit to victims that you are not familiar with their cultural practices, if that is the case, but do not let self-consciousness about being culturally correct distract you from your professional duties.

- Ask victims, for whom you are providing food, what they would like to eat and drink, rather than offering specific food types with which they may be unfamiliar or that may be culturally offensive to them.

- Recognize that the ability to comprehend spoken English is a more easily acquired skill than speaking or writing the language and, therefore, victims who speak with a strong foreign accent, have a limited vocabulary, and who use incorrect English grammar, may still be able to fully understand you.

- Refrain from using token words in victims' native languages, such as "señor/señora" or "gracias," to show a familiarity with their culture. Also, do not improvise with victims' native language or mimic incorrect English grammar to make yourself understood, by saying, for example, "You comprehendo, sï?" or "He no speak English either?"

- Use the word "immigrant" rather than "alien" in referring to victims. Although "alien" may be the correct legal term, it has a negative connotation for some people.

- Find a qualified interpreter/translator if at all possible to communicate with victims who have limited English proficiency. A victim's children, parents, siblings, spouse, relatives, neighbors, or staff at a community ethnic organization are inappropriate substitutes for a professional interpreter. Moreover, their interests may actually be different from or in conflict with those of the victim.

- When communicating through an interpreter, remember that the interpreter is present solely to transmit information back and forth between victims and you, not to explain information or give opinions. Accordingly, speak directly to victims and avoid side conversations with the interpreter as this can cause anxiety, confusion, or mistrust in victims. (See the section on Victims Who Are Deaf or Hard of Hearing in this guidebook for more tips on communicating through an interpreter.)

- Follow these interviewing tips with victims who have limited English proficiency (in the absence of a qualified interpreter), while also being especially vigilant of what your body language is conveying to victims:

  ❑ Speak slowly and pronounce your words clearly, avoiding slang, jargon, and idioms, such as "Howya do'in?"; "Do you want to give a statement or press charges?"; or "How is she holding up?"

  ❑ Listen attentively and patiently.

  ❑ Do not interrupt, correct grammar, or put words in the victim's mouth.

  ❑ Repeat your statements using different words so that you are understood.

  ❑ Use gestures, mime, and props to visually demonstrate your words. When possible, use words that relate to things you both can see.

  ❑ Ask open-ended questions that require more than a yes or no answer and frequently summarize what you understood the victim to say.

  ❑ Do not pretend to understand if you are not sure that you fully understood victims. Instead, repeat to victims what you think you heard them say and ask for clarification.

  ❑ Keep your manner encouraging.

❑ Never raise your voice in an effort to be understood.

❑ Say numerals for street addresses and telephone and other numbers one at a time. For example, give your telephone number, including the area code, as seven, zero, three, eight, three, eight, five, three, one, seven. Do not say: seven, zero, three, eight, three, eight, fifty-three, seventeen.

❑ Allow extra time for communication.

- Empathize with the overriding but unspoken fears of some victims that they or their offender spouse could be deported, or that they could lose custody of their children.

- Understand that victims' immigration status may be legally tied to their offender; i.e., the offender may be the "sponsoring" spouse or employer for the victim's permanent residency petition. It is essential, therefore, that in all cases in which victims' continued legal presence in the United States may depend on the offender, that victims are referred to an immigration law expert; particularly given that violence against women is among the most common types of victimization for immigrants.

- Do not neglect, even with victims who are uncooperative, to use your interview to provide victims with information—as appropriate—on sexual assault, domestic violence, trafficking in persons (see these sections in this guidebook), and other victim-related issues, including the applicable laws and available services. Never lose sight of the possibility that victims may be under duress not to cooperate with you.

- Try to relate to the everyday reality of victims, some of whom live in cultural and linguistic isolation with no extended family or other support network, and who are ineligible for or lack the ability to access support services, or are even unaware the services exist. Never leave victims before connecting them with a local immigrants' rights or refugee resettlement organization, or another social service agency that has staff who are bilingual in English and victims' native language.

- Recognize that victims may not understand or be able to remember the many different types of law enforcement agencies in the United States, such as the sheriff's office, police department, and highway patrol. Be sure to leave your business card with victims so that they will have easily accessible contact information for you and, more important, for your office.

- Do not assume that victims are familiar with telephone answering machines or voice mail. Explain these devices but also follow up with victims, as they may be uncomfortable leaving you a voice message or may not understand how to do so.

- Share with your local immigrant social service agencies the earlier *First Response to Victims of Crime* handbook (2000), which is available in French, Japanese, and Spanish at www.ovc.gov/foreignlang/welcome.html.

SECTION III

FIRST RESPONSE TO SPECIFIC TYPES OF CRIMINAL VICTIMIZATION

Victims of Sexual Assault

Background

Sexual assault is one of the most traumatic types of criminal victimization. In addition to the trauma suffered by victims from their physical injuries, they can also be intensely traumatized by feelings of humiliation at their violation and by the frightful realization that they could have been even more severely injured or killed.

The three primary responsibilities of law enforcement in sexual assault crimes are to (1) protect, interview, and support the victim; (2) collect and preserve evidence that can assist in the apprehension and prosecution of the offender; and (3) investigate the crime and apprehend the offender.

In the investigation and prosecution of sexual assaults, the participatory role of the victim can be even more important than in other crimes because the victim is usually the sole witness to the crime. However, victims are often reluctant to contact law enforcement. They may worry that officers will not believe them, and they may fear retaliation from the offender, particularly when the offender is a family member or acquaintance of the victim, as is often the case.

Further complicating the first response to victims of sexual assault, and the investigation and prosecution of this crime, is that some offenders use alcohol, marijuana, and other memory-erasing drugs to commit a drug-facilitated sexual assault. Whether the drugs are

First Response to Victims of Crime

taken unknowingly—after the offender slips a "date rape" drug into the victim's drink—or voluntarily by the victim for recreational purposes, the drugs render the victim unable to legally consent to sexual acts. Although alcohol is the most commonly used drug in sexual assaults, other drugs used include antihistamines, barbiturates, benzodiazepines such as Rohypnol, and GHB. These drugs are powerful, fast-acting, central nervous system depressants that can render a person physically and mentally incapacitated as well as induce amnesia, whereby acts that occur while the drugs are in effect are not remembered. Victims may not report the assault until days later, in part because the drugs impair their memory and in part because they may not recognize physical signs of the assault. Because these drugs can be detected in a person's body only for a short time, by the time victims notify law enforcement of the assault, the toxicological aspects of the investigation may already be hampered.

The potentially complicated circumstances in sexual assault crimes, and the complexity of feelings experienced by victims, make the approach of a first responder all the more critical. Your initial contact with victims can significantly affect when and if they start down the road to recovery.

Tips on Responding to Victims of Sexual Assault

- Do not be surprised or confused by the nature of victims' emotional reaction to the assault. Be unconditionally supportive and permit victims to express their emotions, the range of which can include crying, angry outbursts, withdrawal and difficulty talking about the assault, screaming, and laughing (which may be in response to stress and relief at having survived the assault).

- Avoid interpreting a victim's calmness or lack of emotional reaction as indicating that a sexual assault may not have occurred or that the victim has not been hurt or traumatized. The victim could be in shock.

- Be attentive to victims' sense of personal dignity and make sure that their bodies are appropriately covered and not immodestly displayed. For example, during your interview of victims, do not lose sight of the possibility that they may be preoccupied with their dirty face, disheveled hair, torn clothing, or otherwise "undignified" appearance. Provide

First Response to Specific Types of Criminal Victimization

victims with covering while being careful not to compromise the collection of any forensic evidence.

- Interact with victims in a calm, professional manner. Exhibiting your outrage or disgust at the crime may cause victims even more trauma. Also, do not make promises that you cannot keep by saying, "We're going to catch whoever did this," or "We'll see that he gets the longest sentence possible."

- Ask victims if they would like you to contact a family member or friend.

- Offer to contact a victim services or sexual assault crisis counselor. Ask victims if they would prefer a male or female counselor. Similarly, ask victims if they are comfortable talking with you and, if not, whether they would prefer a male or female officer.

- Be careful not to come across as overprotective or patronizing. Empower victims by offering your recommendations of treatment or other services as options for victims themselves to consider and decide upon.

- Remember during your interview that it is natural for victims to want to forget, or actually to forget, details of the crime that are difficult for them to accept.

- Establish rapport with victims and explain in advance of your interview that some of your questions may seem irrelevant, strange, or even as if you are blaming them for the assault. Reassure victims that they are not to blame and that your questions are necessary and intended only to gather information to help in the apprehension and successful prosecution of the offender.

- Be mindful of the personal and privacy concerns of victims. They may have a number of concerns, including the possibility that they have been impregnated or have contracted a sexually transmitted disease such as HIV, the virus that causes AIDS; the reactions of their spouse, intimate partner, or parents; media publicity that may reveal their victimization to the public; and the reactions or criticism of neighbors and coworkers if they learn of the sexual assault.

- In cases of a suspected drug-facilitated sexual assault, ask victims if they know what types of alcohol they drank and/or other drugs they took; how many drinks and/or pills they ingested, as well as the drink size (ounces) and pill dosage (milligrams); and, finally, at what times they drank and/or took the pills. Document this information in your report.

- In cases of a suspected drug-facilitated sexual assault, be aware that drugs metabolize at different rates and some are detectable in the blood and urine for only a few hours; thus, specimens need to be collected from victims as soon as possible. Also, it is best for forensic screening if the urine specimen collected is the victims' first urination after the ingestion of the drugs and if victims do not eat or drink anything until after the forensic examination.

- Offer to transport victims to a hospital for medical care and to be seen by a sexual assault nurse examiner (SANE) or forensic medical examiner. The hospital should be one that has a SANE program, where there is one available, and should be in the same state in which the assault occurred. Victims may be physically numb from the emotional shock of the assault, and they need to be examined for physical injuries that may not be easily visible. In addition to requiring treatment, the injuries may provide evidence of the offender's use of force. Understand that victims may feel embarrassed and humiliated that their bodies were exposed during the sexual assault and must be exposed again during the examination. Explain generally what will take place in the examination, including an evaluation of the risks of having contracted a sexually transmitted disease or becoming pregnant (and that medications exist to prevent both); the taking of blood and urine specimens; and the collection of other physical evidence for DNA profiling, to aid in the apprehension and prosecution of the offender. Finally, do not enter the examination room. Medical staff will collect the proper evidence and maintain the chain of custody.

- Notify the hospital of incoming victims/patients and request a private waiting room. If no crisis intervention counselor or victim advocate is available, wait at the hospital until victims are released or have hospital staff notify you before they process the release so that you can escort victims to their final destination. Provide victims with the telephone number and brochure of the local rape crisis center.

- Interview victims with extreme sensitivity and keep to a minimum the number of times they have to recount details of the assault. Consider doing only a preliminary interview, and then immediately transporting victims to the hospital for evaluation, after which you can conduct a more thorough interview.

- Ask victims to give you multiple contact names with telephone numbers and addresses because victims may later decide to stay with relatives or friends rather than in their own home. Also, find out who is most likely to know victims' whereabouts at all times and obtain permission to contact that person, if necessary.

- Offer to answer any further questions that victims may have and provide any further assistance they may need.

Victims of Domestic Violence
Background

Domestic violence is a crime, not a private or family matter, and should be responded to as a crime by law enforcement. Between 1976 and 2004, approximately 19 percent of all homicides in the United States were committed within families or intimate relationships.[33] Moreover, about one-third of female murder victims in recent years were killed by a current or former spouse or a boyfriend or girlfriend.[34] Finally, among women in a 1995–1996 national survey who reported having been raped, physically assaulted, or stalked since they were 18 years old, 64 percent were victimized by a current or former husband, cohabiting partner, boyfriend, or date.[35]

The three primary responsibilities of law enforcement in domestic violence cases are to (1) provide physical safety and security for victims; (2) assist victims by coordinating their referral to support services; and (3) arrest the domestic violence offender as required by law.

Unlike most other crimes, domestic violence is usually not an isolated, sudden, and unexpected incident. It can involve years of emotional trauma, physical injury, and threats to victims' lives, the incidence of which can often become more severe and frequent over time. To the officer who has responded numerous times to the same domestic violence scene, the efforts of first responders to help the victim cope with and recover from the victimization may seem to be of limited consequence. This is not, however, the case. Your appropriate response can have a significant, incremental impact in the sometimes lengthy process of victims' recovery.

Tips on Responding to Victims of Domestic Violence

- Arrive on the scene with a partner if possible; domestic violence calls present potential dangers to responding officers. Introduce yourselves and briefly state that you were dispatched to this address because of a possible injury. Request permission to enter the residence to take your report.

- Separate the parties involved in the domestic violence before interviewing them, even if there is no ongoing violence or argument when you arrive.

- Ask victims if they would like you to contact a family member or friend.

- Do not judge victims or personally comment on their situation. Abusive relationships continue for many reasons. What may seem like good, logical advice offered at the crime scene may actually be ill-advised given the circumstances. For example, advising victims simply to walk away from their situation is not good advice. Many domestic assaults occur as victims are trying to end the relationship.

- Do reassure victims that the domestic violence is not their fault.

- Be aware that domestic violence can be perpetrated by and against both males and females, and that it occurs in same-sex relationships, across all socioeconomic classes, and throughout the entire lifespan, including among older couples.

First Response to Specific Types of Criminal Victimization

- Inquire whether there are children in the family and, if so, determine their whereabouts. Keep in mind that children sometimes hide or are hidden in these circumstances.

- Approach children with care and kindness. Interview them away from their parents' sight and hearing. Look for signs of emotional trauma or distress, and observe the children for any physical indications of abuse. Child abuse is sometimes linked with domestic violence.

- Know your state's laws on domestic violence, including any provisions on mandatory arrest of the "primary" or "predominant aggressor." Also, know the federal laws on domestic violence, including the rights of victims and provisions on gun possession by offenders and persons subject to a protection order.

- Encourage the parties to separate at least overnight in situations where probable cause is lacking for an arrest to be made. If the safety of the party whom you consider to be the victim can be ensured, recommend that the other party leave the home. Traditionally, law enforcement has suggested that victims leave the home, but this approach serves to empower the other party and further disrupt victims' lives, especially when children are involved.

- Offer to temporarily hold in safekeeping any firearms and other weapons at the residence—whether or not an arrest is made and where no legal authority exists to seize the weapons—and strongly urge all parties to voluntarily hand over any weapons to you as a precautionary measure.

- Emphasize to victims that the purpose of your intervention is to help address the problem, not to make the situation worse.

- Advise victims about the criminal nature of domestic violence and the potential for the violence to escalate.

- Be sure to complete a thorough report—photographing and documenting any evidence of injury to victims and damage or disruption at the scene—even if no arrest is made.

- Provide victims with referral information, in writing, on programs for battered women or men, as well as information on domestic violence shelters. This should be done away from offenders.

Victims of Drunk Driving Crashes

Background

In the year 2004, 16,694 people were killed in alcohol-related driving crashes—an average of one person killed every 31 minutes.[36] An estimated 248,000 persons were injured in crashes in which law enforcement reported that alcohol was present—an average of one person injured approximately every 2 minutes.[37] Finally, about 20 percent of all drivers involved in fatal crashes had a blood alcohol concentration of 0.08 or higher,[38] making the act of driving drunk a violent and criminal act, not an accident.

Drunk driving victimization is generally severe and its consequences are long-lasting.[39] Research funded by the National Institute of Mental Health concluded that 5 years after victimization, most persons remain psychologically, physically, and financially impaired.[40] Twenty percent of victims feel they will never again experience a normal life.[41]

The law enforcement officer who is knowledgeable about the unique nature of injury and death in drunk driving crashes will be forever remembered by victims—or the survivors of victims—as a first responder who made a difference. Just always remember, it could have been you or your loved ones who were killed or injured by the drunk driver. Acknowledging this reality will give you patience, compassion, and empathy.

Tips on Responding to Victims of Drunk Driving Crashes

- Avoid comments that disregard or diminish victims' emotional and physical trauma. For example, do not say "You're lucky to be alive" or "At least the driver wasn't speeding." Victims are probably not feeling lucky and such comments are more likely to anger or hurt victims than to provide comfort.

First Response to Specific Types of Criminal Victimization

- Encourage all victims to get immediate medical attention even when no signs of injury are present. Drunk driving crashes are a leading cause of traumatic brain injury (also known as closed head injury), in which the brain is injured without the skull being fractured. Victims with such an injury may show no immediate symptoms and interact normally with first responders. Later, when consequences of the brain injury develop, victims and doctors—in the absence of a medical examination at the time of the crash—may not connect these health problems back to the crash.

- Make sure that your professional attitude and choice of words reflect the fact that drunk driving is a crime—not an accident—with consequences as devastating as those of other violent crimes.

- Understand that victim-drivers may be guilt-ridden at not having avoided the crash when passengers in their vehicles are injured or killed. Gently urge the drivers to confront these emotions with rational thinking. Even when the victim-drivers might possibly have avoided the offender by a particular maneuver, explain to the drivers that their last-second actions were only a small part of a complex sequence of events that led up to the crash.

- Be prepared for intense emotional reactions by victims and even hostility toward you. Victims may believe that law enforcement does not take drunk driving seriously enough, and they may try to argue with you about their views. Remain nonjudgmental, empathetic, and polite, accepting victims' reactions and listening attentively, no matter what their viewpoints. Never contradict or argue with victims.

- Anticipate that victim-passengers in the drunk driver's vehicle may have mixed feelings and make conflicting statements to you about the crash. It can be difficult for them to blame the drunk driver, especially when the driver is a family member or friend. In addition, passengers may be reluctant to provide information to you that could lead to criminal charges being brought against the drunk driver.

- Be supportive of surviving family members who want to view or be with the body of their loved one killed in the crash. Family members often have a strong psychological need to get to the body of their loved one as soon as possible. Your initial inclination may be to deny access to the body out of compassion for the family. Knowing that death from a driving crash almost always entails violent injury to the body, you naturally want to spare families the pain such devastating images will cause them. However, refusing access to their loved one's body will only increase the suffering of surviving family members. Instead, first offer to view the body and describe it in detail to them. If they still want to see and be with the body, support their right to do so. Touching and holding their loved one's body at the scene may be the last opportunity survivors will have to say goodbye while the body is still in its natural state, i.e., before mortuary embalming. Viewing the body can also help surviving family members begin the process of accepting the death of their loved one.

- Choose your words with care and sensitivity. For example, the distinction between "died" and "killed" can take on important significance for surviving family members after a drunk-driving crash fatality. The word "died" ignores the victimization. "Killed" signifies the deliberate or reckless taking of life.

- Look for and place in safekeeping any personal property of victims, such as clothing and jewelry, found at the crash scene. In one survey on satisfaction with the criminal justice system's response to drunk driving crashes, nearly two-thirds of respondents were satisfied with law enforcement's investigation of the case, but many felt that officers had failed to protect the personal property of their loved ones.[42] This perception was a source of hurt and bitterness.

- Review the next section in this guidebook, Survivors of Homicide Victims, for additional tips on responding to the needs of survivors of victims killed in drunk driving crashes.

Survivors of Homicide Victims

Background

Homicide is a crime with more than one victim. Survivors of homicide victims can include parents, a spouse or intimate partner, children, siblings and other next of kin, and sometimes friends, neighbors, and coworkers. Nothing could ever prepare survivors for the day they are notified that their loved one has been killed. They may go into shock upon suddenly learning of the loss of their loved one; react in disbelief and denial; or be overcome with anger that their loved one did not have to die. Survivors' trust in the world, spirituality, and beliefs about social order and justice can be devastated.

Many survivors of homicide victims say that the most traumatic incident of their lives was when they were notified of their loved one's death. Furthermore, an inappropriate notification by first responders can complicate and delay survivors' recovery and prolong their grieving process for years. Death notifications are among the most difficult of duties that law enforcement must perform. However, a properly conducted notification, as outlined below, can help survivors to maintain or regain their trust and belief systems, and aid them in rebuilding their lives.

Tips on Responding to Survivors of Homicide Victims

- Find out as much information as time permits about the survivors of the homicide victim before making notification. Notify the appropriate closest survivor first.

- Make notifications in person.

- Have information to confirm the homicide victim's identity in the event of disbelief or denial by the survivors.

- Refer to the homicide victim by name, out of respect for the victim and survivors. Do not use terms like "the deceased" or "the victim," and do not refer to the victim's body as his or her "remains."

First Response to Victims of Crime

- Familiarize yourself with the details surrounding the homicide victim's death before you conduct the notification. Survivors often want to know the exact circumstances of their loved one's death, including specific information about injuries to the loved one. Answer as many of the survivors' questions as possible, using plain and factual language. By providing information about the victims' injuries, you will have prepared the survivors if they are later called upon to identify their loved one's body or if they hear these details presented in court.

- Be prepared for questions from survivors regarding the whereabouts of the perpetrator, the status of the investigation, and dangers to their own safety.

- Do not bring personal articles of the homicide victim with you to the notification.

- Conduct notifications with a law enforcement or civilian partner. You can locate volunteers who are specially trained in death notification through your local clergies and victim service agencies. When you know that a survivor has a serious health condition, you should take medical personnel with you.

- Have one person take the lead in conducting notifications. The other person should monitor survivors for reactions that may be dangerous to themselves or others. A civilian partner trained in death notification can be especially helpful in counseling the survivors regarding what they may experience or encounter next, both in their personal lives and from the criminal justice system; in sensitively responding to the survivors' questions; and in providing emotional comfort to the survivors, including, for example, a hug or other physical gesture that may go against agency protocol for officers.

- Conduct notifications in a private place—not on the doorstep—after you and the survivors are seated. If a survivor does not sit down, your partner should remain standing close by, in case the survivor collapses.

- Determine if young children are present in the household or other site of notification. Separate adult survivors (or all but one) from the children during notification. Offer to assist adult survivors in telling the children and in getting a babysitter.

- Avoid engaging in smalltalk when you arrive at the survivors' home or other site of notification. Do not build up slowly to the reason for your presence or to the actual announcement of the death of the survivors' loved one. Also, do not use euphemisms for the death, such as "She passed away" or "We lost him." Be compassionately direct and unambiguous in how you notify survivors. For example, "We've come here to tell you something very terrible. Your daughter was killed around 2:30 today during a robbery in the parking lot at the Town Mart Shopping Center. I'm so sorry."

- Be sensitive to the possibility that the homicide victim may have led a lifestyle unbeknownst to the survivors—one that survivors might have disapproved of. Remain strictly nonjudgmental of homicide victims in your notification. Do not state that the homicide was drug-related or related to any other potentially objectionable conduct by the victim. Such comments appear to blame the victims or hold them partially responsible for their own killing. Also, do not use phrases like "being in the wrong place at the wrong time," which can imply that the homicide was accidental.

- Accept survivors' reactions—no matter how passive or intense—in a nonjudgmental and empathetic manner. Survivors may sit quietly and expressionless, or they may cry and scream hysterically.

- Be prepared for the possibility that the survivor will be hostile toward you as a representative of law enforcement and refrain from responding defensively or impolitely.

- Show empathy for survivors' personal pain, but do not say "I understand." You cannot possibly know the individual dynamics of survivors' suffering when their loved one is killed.

- Ask survivors if they would like you to make in-person notification to other close family members.

- Call the more distantly related survivors of the homicide victim at the request of the immediate survivors. Try to arrange for someone to be with these survivors before making your call, and tell them to sit down before conducting your notification. When you notify people by telephone, be especially attentive to your tone of voice, choice of words, and the pace of your speech. After the notification, ask the survivors to allow you to call a family member, friend, or victim advocate to be with them. Tell each survivor whom you call the names of the other persons who have been notified.

- Show respect for survivors' personal and religious or nonreligious beliefs about death. Do not impose your beliefs by saying of the victim, for example, "He's in a better place now."

- Emphasize to survivors that everyone grieves differently and encourage them to be understanding and supportive of each other.

- Avoid the natural inclination to make promises to survivors that you cannot keep by saying, "We're going to catch whoever did this" or "We'll see that the perpetrator receives the longest sentence possible."

- Be sure that survivors have the location name, address, and telephone number for where their loved one's body is being kept and explain the procedures for identifying a loved one's body. Also, offer or arrange transportation to take survivors to this location, as well as back to their homes.

- Let survivors know whom to call if they have questions and provide telephone contacts for support services, including death scene cleanup services, if appropriate. Due to the trauma and confusion that surround death notification, survivors may be unable to fully comprehend or remember much of what you say. Give them written copies of all the information you discuss about support services.

- Finally, make sure that survivors have someone who can stay with them after you leave.

Victims of Human Trafficking

Background

Trafficking in people for prostitution and forced labor is one of the fastest growing areas of international criminal activity.[43] It is a leading source of profits for organized crime, coupled with drugs and weapons.[44] Human trafficking is a modern-day form of slavery whose perpetrators force, defraud, or coerce their victims into labor or commercial sexual exploitation.* According to U.S. Government estimates, between 600,000 and 800,000 persons[45]—about 80 percent of whom are women and girls,[46] and up to 50 percent of whom are minors[47]—are trafficked across international borders worldwide each year, including as many as 17,500 of whom are trafficked to the United States.[48] These numbers do not count the millions of persons trafficked within their own national borders,[49] including U.S. citizens trafficked within the United States.

Victims of human trafficking can be difficult to identify. They may look no different from the other employed persons you may see daily, including motel/hotel maids and private housekeepers or nannies; factory and sweatshop workers; office janitors; migrant farm laborers; and restaurant servers and kitchen help. Nor at first glance will victims look different from other suspects you may have encountered and arrested for prostitution in massage parlors, strip clubs, and modeling, escort, and matchmaking agencies. But trafficking victims are different. They are forced, defrauded, or coerced into these jobs. And contrary to popular belief, victims of human trafficking can be found in small and rural communities throughout America, not just in big U.S. cities, coastal states, or localities with large concentrations of immigrants.

---

*TRAFFICKING VICTIMS PROTECTION ACT OF 2000, Public Law 106-386—October 28, 2000, §103. DEFINITIONS. (8) SEVERE FORMS OF TRAFFICKING IN PERSONS.— The term "severe forms of trafficking in persons" means— (A) sex trafficking in which a commercial sex act is induced by force, fraud, or coercion, or in which the person induced to perform such act has not attained 18 years of age; or (B) the recruitment, harboring, transportation, provision, or obtaining of a person for labor or services, through the use of force, fraud, or coercion for the purpose of subjection to involuntary servitude, peonage, debt bondage, or slavery. (9) SEX TRAFFICKING.— The term "sex trafficking" means the recruitment, harboring, transportation, provision, or obtaining of a person for the purpose of a commercial sex act.

Law enforcement efforts to detect and assist victims of human trafficking can be hampered further because victims often are unable or are afraid to report their situation or to be truthfully responsive when interviewed by officers. Victims may be unwilling to talk openly with you due to traffickers' retaliatory threats to the victims or their families; recriminations from friends and family for their "selfishness" in jeopardizing the chances of others to immigrate to and work in America; distrust of law enforcement and suspicion of officers' collusion with traffickers; and fear of their own imprisonment or deportation. Individuals who have been trafficked for commercial sexual exploitation also fear the stigma of prostitution and the potential loss of their reputations and those of their families if the victims' situation becomes known. Finally, victims may have been instructed by their traffickers to act defiantly during the interview; and, having perhaps been interviewed many times previously by law enforcement (especially victims in the sex industry), they may not come across as submissive or afraid. Thus, unless officers are knowledgeable about the overall dynamics of human trafficking, victims will not be identified and helped, and traffickers will not be caught and prosecuted.

To begin with, officers need to understand the difference between human trafficking and smuggling. Smuggling is the procurement or transport for profit of persons "with their consent" for illegal entry into a country. This facilitation of entry is not, by itself, trafficking. The key distinction between human trafficking and smuggling is that trafficking involves fraud, force, or coercion to induce a commercial sex act or labor. Traffickers usually prey on the desperately poor or otherwise vulnerable (including runaways and physically or mentally challenged persons), luring their victims with fraudulent promises of good jobs and better lives, and then forcing them to work under often brutal and inhumane conditions. It is also important to note that anyone can be a victim of traffickers, including U.S. citizens, legal immigrants, and even persons who consent to being smuggled into the United States, if they are defrauded, forced, or coerced into labor or a commercial sex act.

First Response to Specific Types of Criminal Victimization

DIFFERENCES BETWEEN HUMAN TRAFFICKING AND SMUGGLING[50]

TRAFFICKING		SMUGGLING
Must Contain an Element of Force, Fraud, or Coercion (actual, perceived, or implied), unless under 18 years of age involved in commercial sex acts.		The person being smuggled is generally cooperating.
Forced Labor and/or Exploitation.		There is no actual or implied coercion.
Persons trafficked are victims.		Persons smuggled are violating the law. They are not victims.
Enslaved, subjected to limited movement or isolation, or had documents confiscated.		Persons are free to leave, change jobs, etc.
Need not involve the actual movement of the victim.		Facilitates the illegal entry of person(s) from one country into another.
No requirement to cross an international border.		Smuggling always crosses an international border.
Person must be involved in labor/services or commercial sex acts, i.e., must be "working."		Person must only be in the country or attempting entry illegally.

This chart is not intended to provide a precise legal distinction of the differences between smuggling and trafficking. Instead, the chart illustrates general fact scenarios that are often seen in smuggling and trafficking incidents. The specific fact scenarios in trafficking and smuggling incidents are often complex, and in such cases expert legal advice should be sought.

Gaining the trust of suspected victims of human trafficking is essential to identifying victims of this crime and to assisting them. Being on the lookout for certain clues, and knowing some of the questions to ask, can also aid in your detection of victims. A few of the clues that, taken together, may indicate that you have come upon a case of human trafficking are listed below.

You should suspect human trafficking when the individual you are interviewing—

- Is accompanied by another person who seems controlling (possibly the trafficker).

- Does not speak on his or her own behalf, or does not speak English.

- Is not in personal possession of a passport or other forms of identification.

- Appears submissive, or seemingly fearful or depressed.

- Shows signs of physical or psychological abuse.

Furthermore, asking questions such as those listed below—although they should not be used as a definitive checklist—can help you determine if an individual is a victim of trafficking:

- What type of work do you do?

- Can you leave your job or change your residence if you want to?

- Do you owe money to your employer? What would happen if you didn't repay that money?

- What are your working and living conditions like?

- Can you come and go as you please?

- Do you have to ask permission to eat, sleep, go to the bathroom, or make a telephone call?

- Are there locks on your doors or windows so you cannot get out?

- Have you or your family members been threatened?

- Are you afraid of anyone? Why?

- Have your identification papers or other documents been taken from you?

First Response to Specific Types of Criminal Victimization

Finally, when a law enforcement agency is planning to bust a suspected human trafficking operation, the agency's contacts and cooperation prior to the bust with a local, if possible, or national, nongovernmental organization (NGO) that focuses on providing services to victims of trafficking can be very helpful both to the victims and the first responders. Some victims report that they were as traumatized by their arrest as criminal suspects as they were by their trafficking experience. NGO representatives can help alleviate this trauma by assuring victims that (1) the NGO is not the government; (2) its staff do not arrest or deport people; (3) the first and foremost concern of the NGO is for the safety of victims; (4) trafficking victims have legal rights; and (5) victims who assist law enforcement may be eligible for social services and other benefits. Such intervention by NGOs during the interview process may persuade victims who may initially be hesitant to speak up to eventually talk with officers.

In conclusion, it is local law enforcement and other community-based entities, rather than federal agencies, that are most likely to first encounter victims of human trafficking. And, while trafficking in humans is largely a hidden crime, trafficking victims are in plain sight if you, as the first responder, know what to look for.

Tips on Responding to Victims of Human Trafficking

- Look beneath the surface. The victim you have encountered, or suspect you are questioning regarding a criminal offense, such as prostitution, may be a victim of human trafficking. Call the U.S. Department of Justice's Trafficking in Persons and Worker Exploitation Task Force Complaint Line at 1-888-428-7581 and the U.S. Department of Health and Human Services' Trafficking Information and Referral Hotline at 1-888-373-7888 to report trafficking crimes and access supportive services if you encounter an individual who may be a victim of trafficking. Operators have access to interpreters who can speak with you and with victims in their own language.

- Keep in mind that any criminal acts committed by trafficked persons may have been committed as a result of force, fraud, or coercion. Undocumented persons, for example, may have come to the United States believing their traffickers' fraudulent promises that they would be able to obtain legal documents once here. Or persons may have entered the United States legally but are now illegally present because traffickers withheld their documents and thereby allowed their visas to expire.

- Go slowly, be patient, and allow for more time in your interview of suspected trafficking victims who are immigrants. Also, review the section in this guidebook on Immigrant Victims.

- Respect a victim's dignity in your interview. Never ask the question, "Are you a slave?" Such questions may be so humiliating that they elicit victims' denial of the reality of their situation. Instead, ask questions like those provided earlier that concentrate on restrictions to the individual's freedom and the ability to come and go as one pleases.

- Realize that victims will likely not see themselves as victims of a crime or know that their treatment by traffickers is against the law. They may instead see themselves as persons enduring temporary suffering and hardship for the sake of a better life in the future; or, they may even believe that their own actions brought about the traffickers' abuse.

- Be alert to behavior of victims that is symptomatic of the "Stockholm Syndrome." This syndrome is a psychological response sometimes exhibited by hostages. The hostage victims respond to their situation by the defense mechanism of identification with their captors. Victims can become emotionally and sympathetically attached to their captors and loyally defend them.

- Separate suspected victims from all other persons accompanying them before interviewing each victim individually. These other persons could be the traffickers posing as a victim's spouse, family member, legitimate employer, or even coworker and fellow victim. Do not even interview persons whom you have confirmed to be victims in front of each other. Victims may not feel safe divulging information that puts them at risk in front of others, even when the others are victims themselves.

- Try to use a female interpreter who is not local or affiliated with law enforcement for your interview of trafficked women. Make sure that the

First Response to Specific Types of Criminal Victimization

interpreter is qualified, has been screened for confidentiality, and is in no way connected to the traffickers.

- Do not say to victims, "Why didn't you just leave?"

- Treat trafficking victims with the same level of professionalism and compassion as other victims of crime regardless of their immigration status or the situation in which you find them.

- Tell victims that you want them to be safe and protected from the people who hurt them. Let victims know that the U.S. Government can help them and may also be able to assist in the reunification of victims with their family members.

- Expect victims' statements to change as their trust of you develops.

- Be cautious of anyone's attempt to contact suspected victims even after they are in the "safe" custody of law enforcement. Traffickers sometimes hire attorneys to "represent" victims arrested for prostitution, or hire other agents from the victims' ethnic community to intimidate them and influence their statements.

- Be aware that victims may have been drugged by traffickers and forced into addictions that will not disappear overnight.

- Facilitate medical attention for victims. All victims will have basic health care needs and most will require emergency care for serious physical, psychological, and possibly sexual abuse. Posttraumatic stress symptoms are common among victims as are intense feelings of guilt, fear, shame, anger, depression, disorientation, betrayal, and distrust.

- Explain to victims what their medical attention may involve while you transport them to a clinic or hospital, or while you wait with them in the emergency room. This may be the first medical examination the victims have ever had, so explain in reassuring terms the basic procedures they can expect (intake questions/medical history, physical examination for and treatment of injuries, collection and testing of blood and urine specimens, prescriptions for medicine).

- Understand that even after victims have been rescued from their traffickers, they generally are incapable of finding support services due to the isolation they suffered while in captivity. An extensive network of culturally and linguistically appropriate service providers is required to meet victims' urgent and acute needs. This can especially affect first responders, who may, themselves, have to initiate such support for victims.

- Connect victims as soon as possible with a local or national organization that offers assistance services, including advocacy and legal representation, to victims of trafficking. These organizations can educate victims about their rights and responsibilities; assist in providing for their basic needs; and help to reduce their anxiety and fear, particularly toward law enforcement. See www.ovc.gov/help/traffickingmatrix.htm for a list of organizations with programs to help victims of trafficking.

- Familiarize yourself with the federal Trafficking Victims Protection Act of 2000 (TVPA) and subsequent TVPA Reauthorization Acts, and with your pertinent state trafficking laws, if any. Consider other laws, in addition to trafficking in persons' legislation, that may apply to a case. Under TVPA, victims who are not U.S. citizens may be eligible for special visas and other benefits and services, including the following: emergency medical assistance, food and shelter, counseling, and legal and immigration advice. U.S. citizens who are trafficking victims already have access to many federal and state benefits.

- Distribute the brochure *Information for Victims of Trafficking in Persons and Forced Labor* to suspected victims. This brochure, from the U.S. Department of Justice and the U.S. Department of State, is available in English at www.usdoj.gov/crt/crim/wetf/victimsbrochure.pdf, and in traditional Chinese, Korean, Spanish, Thai, and Vietnamese at www.ovc.gov/foreignlang/welcome.html. For service providers and other community-based organizations, provide the companion brochure *Trafficking in Persons: A Guide for Non-Governmental Organizations* at www.usdoj.gov/crt/crim/wetf/trafficbrochure.pdf. Also, fact sheets on issues such as sex trafficking, labor trafficking, child victims of human trafficking, and federal efforts to assist victims are available in English, traditional Chinese, Polish, Russian, and Spanish from the U.S. Department of Health and Human Services at www.acf.hhs.gov/trafficking.

Victims of Mass Casualty Crimes

Background

Mass casualty crimes in the United States, whether they are acts of workplace or school violence, domestic or international terrorism, or other acts involving the criminal victimization of multitudes of people, are infrequent and highly abnormal incidents. They are, however, no longer unimaginable to anyone; nor can law enforcement afford not to anticipate and try to prepare for them.

Even more so than other crimes, a mass casualty crime presents unique challenges for law enforcement, specific to the incident itself and its victims. For example, the sheer number of victims can overwhelm emergency management resources. In addition, officers may find themselves responding to victims in horrific surroundings of human injury, death, and material destruction, and under dangerous conditions of possible further harm from known or unknown sources. Finally, although every criminal act can have devastating consequences for its victim, a mass casualty crime is a devastating act that may include in its wake not only a large number of individual victims, their families, friends, and coworkers, but also whole communities and the entire Nation.

Yet, as we have seen from past incidents of terrorism and other mass casualty crimes, the public safety personnel involved in the first response to victims—no matter how brief or hurried the personal interaction between responders and victims—have the profound opportunity to reaffirm society's faith in the goodness of humankind while positively influencing the emotional recovery of an individual victim.

The guidelines presented here are by no means comprehensive; nor are they practical in, or appropriate for, every situation. They are, however, basic considerations for officers trying to respond compassionately and most effectively to victims amid the competing duties of securing the scene and imposing order in the chaos of a mass casualty crime.

Tips on Responding to Victims of Mass Casualty Crimes

- Identify immediately those physically and psychologically injured victims who need urgent medical care and/or mental health crisis intervention. Behavioral indicators of psychological injury can include hyperventilating; trembling or "freezing"; odd body movements; agitation; emotional numbness, as evidenced by fixed staring or a marked lack of expression and body movement; rambling speech; loud wailing; or an inability to comprehend one's situation and surroundings.

- Display as much calm assurance as you can to reduce panic in victims.

- Begin helping victims reorient themselves immediately after you rescue them—they have just lost the psychological connection to the world with which they are familiar—by identifying yourself and your role to victims, and projecting support and compassion in your interactions with them.

- Help victims recognize that the danger has passed by saying "It's over now, you're safe"; "We're getting you out of here now, it's okay"; or "You're going to be all right, we're taking you to safety now."

- Give polite and kind but firm instructions to victims, who may be stunned and immobilized by the incident—or, conversely, unproductively overactive—and not know what they should do next. Direct victims who are able to walk, away from the site and away from severely injured victims and any continuing danger.

- Understand that when people are confronted with life-threatening situations, they respond with survival-mode functioning, characterized by flight, fight, or freeze behaviors. Symptoms of the flight response are extreme anxiety, avoidance, disbelief, denial, or regressed childlike behaviors. Rage or aggression is symptomatic of the fight response, and apathy, emotional numbness, or aimlessness indicates the freeze response. For example, people in the freeze response may wander aimlessly in the devastation of an incident, thereby placing themselves and emergency responders in further danger. Victims exhibiting such

First Response to Specific Types of Criminal Victimization

flight, fight, or freeze behaviors need your protection until mental health crisis intervention can be provided.

- Do not be surprised by or judgmental of victims who appear joyous, full of pride, or in high spirits. It is not uncommon for people to experience such euphoric feelings immediately after having just survived a life-threatening incident. However, be mindful that, for other persons at the scene who are not knowledgeable about the mind's psychological defense mechanisms to overwhelming stress, these euphoric reactions can be difficult to tolerate in surroundings of death and devastation.

- Protect victims—including those being interviewed or waiting to be interviewed—from exposure to further traumatic stimuli by locating or devising a real or even symbolic area of safe haven for them. The fewer traumatic stimuli victims see, hear, smell, taste, and feel, the better off they will be.

- Remind victims to breathe. When people are traumatized or frightened, they often stop breathing normally. Have victims close their eyes and take deep, slow breaths until they calm down.

- Make sure that victims are not cold and that warm beverages and food are offered to them and that restroom facilities are provided.

- Realize that victims may have difficulty concentrating on your questions, understanding and following your instructions, and remembering the information you provide until their basic physical and psychological needs are met. Also, be aware that traumatic stress can have the following effects on victims' mental functioning: disorientation and confusion, distorted sense of time, slowed thinking, shortened attention span and narrowed ability to focus, indecisiveness, and memory loss.

- Be attentive to victims' dignity and comfort, and try to ensure that their bodies are appropriately covered and not immodestly displayed. For example, do not lose sight during your interview of victims that they may be preoccupied with their dirty or bloody face, torn clothing, or otherwise "undignified" appearance. Provide victims with covering and otherwise try to protect them from any continuing indignities without compromising the collection of forensic evidence.

- Give clear and simple explanations of law enforcement and other emergency response procedures, which are sometimes confusing and distressing to victims.

- Empathize and be extremely sensitive to victims' shock and grief over the possible injury to, or death of, their loved ones, friends, and coworkers. Recognize that victims may need a period of silence from you so they can reflect on and get in touch with their own feelings.

- Be as factually responsive as you can to victims' need for information about their injuries; the condition and whereabouts of other victims; the nature of the incident; rescue and emergency operations underway; and whether the suspected perpetrators have been apprehended. Sharing these informational updates with victims, and your efforts to get answers to their individual questions, will help to reduce victims' anxiety. However, be careful about passing on rumors to victims; avoid giving out information on the identities of other persons at the scene who were killed; and do not tell victims about the death of, or severe injury to, their loved ones until victims are in an appropriate setting and have been stabilized or are under medical supervision.

- Limit intrusions into victims' privacy by onlookers and the news media. However, facilitate communication and provide security for those victims who want to speak with the media. Advise victims that they are free to end an interview with media representatives at any time and for any reason, and provide for their quick and safe escort away from the interview site.

- Assist victims in verifying the personal safety of loved ones as well as in locating them. Reconnect victims with their family, friends, and other trusted people as soon as possible.

- Document the identities of, and contact information for, both the injured and non-injured victims at the scene of the crime. Their families will be contacting law enforcement to determine their safety and whereabouts.

- Provide victims with written information on resources for immediate and long-term help, including your agency's victim services unit, if one exists; the prosecutor's and victim-witness assistance offices; mental health agencies, local crisis intervention centers, and support groups; emergency shelters and food centers; faith-based resources; and the state crime victim compensation program.

- Remember that no matter how frantic, on-the-run, or curtailed the interaction is between first responders and victims, responders can help to restore victims' sense of control over their lives when that interaction is characterized by sensitivity, compassion, and respect.

SECTION IV

DIRECTORY OF NATIONAL SERVICE PROVIDERS

Provide the following national resources and hotlines to victims as appropriate:

Alzheimer's Disease

Alzheimer's Association
1-800-272-3900 (TDD: 312-335-8882)
www.alz.org

Safe Return
Alzheimer's Association
Incident Line: 1-800-572-1122 (TDD: 312-335-8882)
Nonemergency Line: 1-888-572-8566 (TDD: 312-335-8882)
www.alz.org/Services/SafeReturn.asp

Americans with Disabilities Act of 1990 and Section 504 of the Rehabilitation Act of 1973

Americans with Disabilities Act Information Line
U.S. Department of Justice
1-800-514-0301 (TDD: 1-800-514-0383)
www.ada.gov

Office for Civil Rights
Office of Justice Programs
U.S. Department of Justice
202-307-0690 (TDD: 202-307-2027)
www.ojp.usdoj.gov/about/offices/ocr.htm

Blindness or Other Vision Impairments

American Council of the Blind
1-800-424-8666
www.acb.org

American Foundation for the Blind
1-800-232-5463
www.afb.org

Children

Child Abuse Hotline
Bureau of Indian Affairs
U.S. Department of the Interior
1-800-633-5155

Child Welfare Information Gateway
Children's Bureau, Administration for Children and Families
U.S. Department of Health and Human Services
1-800-394-3366
www.childwelfare.gov

Childhelp® USA National Child Abuse Hotline
1-800-422-4453
www.childhelp.org

Covenant House Nineline
1-800-999-9999 (TDD: 1-800-999-9915)
(provides shelter, crisis care, and other services to homeless and runaway youth)
www.covenanthouse.org

Juvenile Justice Clearinghouse
Office of Juvenile Justice and Delinquency Prevention
Office of Justice Programs
U.S. Department of Justice
1-800-851-3420 (TDD: 1-877-712-9279)
www.ojjdp.ncjrs.gov/programs/ProgSummary.asp?pi = 2

National Center for Missing and Exploited Children
1-800-843-5678 (TDD: 1-800-826-7653)
www.missingkids.com

National Children's Alliance
1-800-239-9950
www.nca-online.org

National Runaway Switchboard
1-800-621-4000 (TDD: 1-800-621-0394)
www.1800runaway.org

Deafness or Hard of Hearing

Alexander Graham Bell Association for the Deaf and Hard of Hearing
1-866-337-5220 (TDD: 202-337-5221)
www.agbell.org

National Association of the Deaf
301-587-1788 (TDD: 301-587-1789)
www.nad.org

National Institute on Deafness and Other Communication Disorders
National Institutes of Health
U.S. Department of Health and Human Services
1-800-241-1044 (TDD: 1-800-241-1055)
www.nidcd.nih.gov

Registry of Interpreters for the Deaf
703-838-0030 (TDD: 703-838-0459)
www.rid.org

Telecommunications Relay Services
711
(These services enable callers with hearing and speech disabilities who use text telephones (i.e., TTYs, TDDs) and callers who use voice telephones to communicate with each other through third-party assistance.)

Domestic Violence

Battered Women's Justice Project
1-800-903-0111
www.bwjp.org

Family Violence Department
National Council of Juvenile and Family Court Judges
1-800-527-3223
www.ncjfcj.org/content/view/20/94

National Domestic Violence Hotline
1-800-799-7233 (TDD: 1-800-787-3224)
www.ndvh.org

National Resource Center on Domestic Violence
1-800-537-2238 (TDD: 1-800-553-2508)
www.nrcdv.org

Drunk Driving Crashes

Mothers Against Drunk Driving
1-800-438-6233
Victims Helpline: 1-877-623-3435
www.madd.org/victims

Immigrants

Language Line Services
1-800-528-5888 (credit card required for emergency interpreter/translation service)
1-877-886-3885 (general information and account setup for long-term service)
www.languageline.com

Office of Special Counsel for Immigration-Related Unfair
Employment Practices
Civil Rights Division
U.S. Department of Justice
1-800-255-7688 (TDD: 1-800-237-2515)
www.usdoj.gov/crt/osc

Mass Casualty Crimes

Office for Victim Assistance
Federal Bureau of Investigation
www.fbi.gov/hq/cid/victimassist/home.htm
www.fbi.gov/contact/fo/fo.htm (directory of FBI offices, for locating the nearest victim specialist)

Mental Illness

Depression and Bipolar Support Alliance
1-800-826-3632
www.dbsalliance.org

National Alliance on Mental Illness
1-800-950-6264 (TDD: 703-516-7227)
www.nami.org

Mental Health America Help Desk
1-800-969-6642 (TDD: 1-800-433-5959)
www.nmha.org/infoctr/index.cfm

National Suicide Prevention Lifeline
1-800-273-8255 (TDD: 1-800-799-4889)
www.suicidepreventionlifeline.org

Treatment Advocacy Center
703-294-6001; 703-294-6002
www.psychlaws.org

Mental Retardation

American Association on Intellectual and Developmental Disabilities
1-800-424-3688
www.aamr.org

National Down Syndrome Congress
1-800-232-6372
www.ndsccenter.org

The Arc
1-800-433-5255
www.thearc.org

Older Persons

Eldercare Locator
(links older persons or their caregivers with local senior service agencies and organizations)
Administration on Aging
U.S. Department of Health and Human Services
1-800-677-1116
www.eldercare.gov

National Association of Area Agencies on Aging
202-872-0888
www.n4a.org

Sexual Assault

CDC-INFO
(obtain information anonymously on sexually transmitted diseases and get referrals to clinics)
Centers for Disease Control and Prevention
U.S. Department of Health and Human Services
1-800-232-4636 (TDD: 1-888-232-6348)
www.cdc.gov/std

National Sexual Assault Hotline
Rape, Abuse and Incest National Network
1-800-656-4673
www.rainn.org

National Sexual Violence Resource Center
1-877-739-3895 (TDD: 717-909-0715)
www.nsvrc.org

Survivors of Homicide Victims
National Organization of Parents Of Murdered Children
1-888-818-7662
www.pomc.org

Trafficking in Persons

Office for Victims of Crime-Funded Grantee Programs To Help Victims of Trafficking
http://www.ovc.gov/help/traffickingmatrix.htm

Trafficking in Persons and Worker Exploitation Task Force Complaint Line
U.S. Department of Justice
1-888-428-7581 (voice and TDD)
www.usdoj.gov/whatwedo/whatwedo_ctip.html

Trafficking Information and Referral Hotline
Administration for Children and Families
U.S. Department of Health and Human Services
1-888-373-7888
www.acf.hhs.gov/trafficking

Other National Victim Service Providers and Resources

Bureau of Justice Assistance
Office of Justice Programs
U.S. Department of Justice
202-616-6500
www.ojp.usdoj.gov/BJA

First Response to Victims of Crime

COPS Office Response Center
Office of Community Oriented Policing Services
U.S. Department of Justice
1-800-421-6770
www.cops.usdoj.gov

Families and Friends of Violent Crime Victims
1-800-346-7555 (TDD: 425-355-6962)
www.fnfvcv.org

National Center for Victims of Crime
1-800-394-2255 (TDD: 1-800-211-7996)
www.ncvc.org

National Clearinghouse for Alcohol and Drug Information
Substance Abuse and Mental Health Services Administration
U.S. Department of Health and Human Services
1-800-729-6686 (Spanish: 1-877-767-8432) (TDD: 1-800-487-4889)
http://ncadi.samhsa.gov

National Criminal Justice Reference Service
1-800-851-3420 (TDD: 1-877-712-9279)
www.ncjrs.gov

National Fraud Information Center/Internet Fraud Watch
National Consumers League
1-800-876-7060
www.fraud.org

National Organization for Victim Assistance
1-800-879-6682
www.try-nova.org

Office for Victims of Crime
Office of Justice Programs
U.S. Department of Justice
202-307-5983 (TDD: 202-514-7908)
www.ovc.gov

Office for Victims of Crime Resource Center
Office of Justice Programs
U.S. Department of Justice
1-800-851-3420 (TDD: 1-877-712-9279)
www.ovc.gov/ovcres

Office for Victims of Crime Training and Technical Assistance Center
Office of Justice Programs
U.S. Department of Justice
1-866-682-8822 (TDD: 1-866-682-8880)
www.ovcttac.gov

Office of Juvenile Justice and Delinquency Prevention
Office of Justice Programs
U.S. Department of Justice
202-307-5911
http://ojjdp.ncjrs.gov

Office on Violence Against Women
U.S. Department of Justice
202-307-6026 (TDD: 202-307-2277)
www.usdoj.gov/ovw

SECTION V

ENDNOTES

[1] Bureau of Justice Statistics. (March 1994). *Elderly Crime Victims: National Crime Victimization Survey.* Washington, DC: U.S. Department of Justice, Office of Justice Programs, Bureau of Justice Statistics, NCJ 147002. Retrieved August 21, 2006, from www.ojp.usdoj.gov/bjs/pub/ascii/ecv.txt.

[2] Office for Victims of Crime. (June 1999). *Breaking the Cycle of Violence: Recommendations to Improve the Criminal Justice Response to Child Victims and Witnesses.* (OVC Monograph). Washington, DC: U.S. Department of Justice, Office of Justice Programs, Office for Victims of Crime, NCJ 176983. Retrieved August 21, 2006, from www.ojp.usdoj.gov/ovc/publications/factshts/monograph.htm.

[3] Ibid.

[4] Ibid.

[5] Ibid.

[6] Bureau of Justice Statistics. (June 2006). "Percent of Victimizations Reported to the Police, by Type of Crime and Age of Victims," Table 96, Personal Crimes, 2004. In *Criminal Victimization in the United States—Statistical Tables.* Washington, DC: U.S. Department of Justice, Office of Justice Programs, Bureau of Justice Statistics. Retrieved August 21, 2006, from www.ojp.usdoj.gov/bjs/abstract/cvusst.htm and www.ojp.usdoj.gov/bjs/pub/pdf/cvus/current/cv0496.pdf.

[7] Bureau of Justice Statistics. (June 2006). "Victimization Rates for Persons Age 12 and Over, by Type of Crime and Age of Victims," Table 3, Personal Crimes, 2004. In *Criminal Victimization in the United States—Statistical Tables.* Washington, DC: U.S. Department of Justice, Office of Justice Programs, Bureau of Justice Statistics.

Retrieved August 21, 2006, from www.ojp.usdoj.gov/bjs/pub/pdf/cvus/current/cv0403.pdf.

[8] Marge, Dorothy K. (ed.). (2003). "Executive Summary." In *A Call to Action: Ending Crimes of Violence Against Children and Adults with Disabilities—A Report to the Nation, 2003*. Syracuse, NY: SUNY Upstate Medical University, Department of Physical Medicine and Rehabilitation. Retrieved August 21, 2006, from www.upstate.edu/pmr/marge.pdf.

[9] Kraus, L.; Stoddard, S.; and Gilmartin, D. (1996). "Section 1: Prevalence of Disabilities." In *Chartbook on Disability in the United States, 1996.* (An InfoUse Report). Washington, DC: U.S. Department of Education, National Institute on Disability and Rehabilitation Research, H133D50017. Retrieved August 21, 2006, from www.infouse.com/disabilitydata/disability/intro.php.

[10] Ibid.

[11] National Institute on Aging. (first published March 2002; last reviewed January 2005). "Alzheimer's Disease Defined." In *NIH Senior Health.* Bethesda, MD: U.S. Department of Health and Human Services, National Institutes of Health, U.S. National Library of Medicine, National Institute on Aging. Retrieved August 22, 2006, from www.nihseniorhealth.gov/alzheimersdisease/defined/01.html.

[12] Alzheimer's Association. (2007). *Alzheimer's Disease Facts and Figures, 2007.* Chicago, IL: Alzheimer's Association. Retrieved November 27, 2007, from www.alz.org/national/documents/Report_2007FactsAndFigures.pdf.

[13] Ibid.

[14] Ibid.

[15] Ibid.

[16] Alzheimer's Association. (2006). *Wandering Behavior—Preparing for and Preventing It* (Fact Sheet). Chicago, IL: Alzheimer's Association. Retrieved April 16, 2007, from www.alz.org/documents/national/FSSRCaregivers.pdf.

[17] Hobbs, Frank, and Stoops, Nicole. (November 2002). "Highlights." In *Demographic Trends in the 20th Century* (Census 2000 Special Reports). Washington, DC: U.S. Department of Commerce, Economics and Statistics Administration, U.S. Census Bureau, Series CENSR-4. Retrieved August 22, 2006, from www.census.gov/prod/2002pubs/censr-4.pdf.

[18] National Institute of Mental Health. (updated June 2006). "Statistics." In *Health Information/Mental Health Topics.* Bethesda, MD: U.S. Department of Health and Human Services, National Institutes of Health, National Institute of Mental Health. Retrieved August 22, 2006, from www.nimh.nih.gov/HealthInformation/statisticsmenu.cfm.

[19] Office of the Surgeon General. (1999). "Chapter 2, The Fundamentals of Mental Health and Mental Illness: Epidemiology of Mental Illness." In *Mental Health: A Report of the Surgeon General.* Washington, DC: U.S. Department of Health and Human Services, U.S. Public Health Service, Office of the Surgeon General. Retrieved August 22, 2006, from www.surgeongeneral.gov/library/mentalhealth/home.html.

[20] The Arc. (revised October 2004). *Introduction to Mental Retardation.* Silver Spring, MD: The Arc. Retrieved August 22, 2006, from www.thearc.org/faqs/intromr.pdf.

[21] Luckasson, Ruth. (1992). "Chapter 11, People with Mental Retardation as Victims of Crime." In Ronald W. Conley, Ruth Luckasson, and George N. Bouthilet. (eds.). *The Criminal Justice System and Mental Retardation—Defendants and Victims.* Baltimore, MD: Paul H. Brooks Publishing Company.

[22] The Arc. (revised January 1994). *Employment of People with Mental Retardation.* Silver Spring, MD: The Arc. Retrieved August 22, 2006, from www.thearc.org/faqs/emqa.html.

[23] National Institutes of Health. (May 2006). "Study Finds Most Americans Have Good Vision, But 14 Million Are Visually Impaired." *NIH News.* Bethesda, MD: U.S. Department of Health and Human Services, National Institutes of Health. Retrieved August 22, 2006, from www.nih.gov/news/pr/may2006/nei-09.htm.

[24] Prevent Blindness America and National Eye Institute. (2002). "Section 1: Vision Impairment & Blindness." In *Vision Problems in the U.S.—Prevalence of Adult Vision Impairment and Age-Related Eye Disease in America.* Chicago, IL: Prevent Blindness America. Retrieved August 22, 2006, from www.nei.nih.gov/eyedata/pdf/VPUS.pdf.

[25] National Eye Institute. (modified November 2004). "Low Vision and Blindness Rehabilitation." In *National Plan for Eye and Vision Research.* Bethesda, MD: U.S. Department of Health and Human Services, National Institutes of Health, National Eye Institute. Retrieved August 22, 2006, from www.nei.nih.gov/strategicplanning/np_low.asp.

[26] The Cleveland Clinic. (reviewed March 2002). "Hearing Loss." In *Health Information Center.* Cleveland, OH: The Cleveland Clinic. Retrieved August 23, 2006, from www.clevelandclinic.org/health/health-info/docs/2700/2733.asp?index=10059.

[27] National Institute on Deafness and Other Communication Disorders. (January 2001). "Hearing Loss and Older Adults." In *Health Information.* Bethesda, MD: U.S. Department of Health and Human Services, National Institutes of Health, National Institute on Deafness and Other Communication Disorders, NIH Publication No. 01-4913. Retrieved August 23, 2006, from www.nidcd.nih.gov/health/hearing/older.htm.

[28] Ibid.

[29] McNeil, Jack. (February 2001). *Americans with Disabilities, 1997.* (Current Population Reports). Washington, DC: U.S. Department of Commerce, Economics and Statistics Administration, U.S. Census Bureau, P70-73. Retrieved August 23, 2006, from www.census.gov/prod/2001pubs/p70-73.pdf.

[30] Ibid.

[31] Larsen, Luke J. (August 2004). *The Foreign-Born Population in the United States: 2003.* (Current Population Reports). Washington, DC: U.S. Department of Commerce, Economics and Statistics Administration, U.S. Census Bureau, P20-551. Retrieved August 23, 2006, from www.census.gov/prod/2004pubs/p20-551.pdf.

[32] Office for Victims of Crime. (July 1999). *Office for Victims of Crime International Activities.* (Fact Sheet). Washington, DC: U.S. Department of Justice, Office of Justice Programs, Office for Victims of Crime, FS000229. Retrieved August 23, 2006, from www.ovc.gov/publications/factshts/pdftxt/interact.pdf.

[33] Fox, James Alan, and Zawitz, Marianne W. (revised June 2006). "Intimate Homicide." In *Homicide Trends in the United States.* Washington, DC: U.S. Department of Justice, Office of Justice Programs, Bureau of Justice Statistics. Retrieved August 23, 2006, from www.ojp.usdoj.gov/bjs/homicide/intimates.htm.

[34] Ibid.

[35] Tjaden, Patricia, and Thoennes, Nancy. (November 2000). "Executive Summary." In *Full Report of the Prevalence, Incidence, and Consequences of Violence Against Women: Findings from the National Violence Against Women Survey.* Washington, DC: U.S. Department of Justice, Office of Justice Programs, National Institute of Justice and U.S. Department of Health and Human Services, Centers for Disease Control and Prevention, National Center for Injury Prevention and Control, NCJ 183781. Retrieved August 23, 2006, from www.ncjrs.gov/pdffiles1/nij/183781.pdf.

[36] According to the National Highway Traffic Safety Administration, a motor vehicle crash is *alcohol-related* if at least one driver or nonoccupant (such as a pedestrian or bicyclist) involved in the crash has a blood alcohol concentration (BAC) of 0.01 gram per deciliter (g/dL) or higher. Any fatality that occurs in an alcohol-related crash is considered an alcohol-related fatality. The term "alcohol-related" does not indicate that a crash or fatality was caused by the presence of alcohol.

As of 2004, 45 states, the District of Columbia, and Puerto Rico had created laws making it illegal to drive with a BAC of 0.08 g/dL or higher. Of the 16,694 people who died in alcohol-related crashes in 2004, 14,409 (86 percent) were killed in crashes in which at least one driver or nonoccupant had a BAC of 0.08 g/dL or higher.

National Center for Statistics and Analysis. (2004). *Alcohol.* (Traffic Safety Facts, 2004 Data). Washington, DC: U.S. Department of Transportation, National Highway Traffic Safety Administration, National Center for Statistics and Analysis, DOT HS 809 905. Retrieved August 23, 2006, from http://www-nrd.nhtsa.dot.gov/pdf/nrd-30/NCSA/TSF2004/809905.pdf.

[37] Ibid.

[38] Ibid.

[39] Mercer, Dorothy. (1995). *Drunk Driving Victim Impact Panels: Victim Outcomes.* Washington, DC: U.S. Department of Health and Human Services, National Institutes of Health, National Institute of Mental Health. Cited by Stephanie Frogge and Janice Lord (June 2002; updated December 2004), "Chapter 13: Drunk Driving." In Anne Seymour, Morna Murray, Jane Sigmon, Melissa Hook, Christine Edmunds, Mario Gaboury, and Grace Coleman. (eds.). (June 2002; updated July 2005). *National Victim Assistance Academy Textbook.* Washington, DC: U.S. Department of Justice, Office of Justice Programs, Office for Victims of Crime. Retrieved August 23, 2006, from www.ojp.usdoj.gov/ovc/assist/nvaa2002/welcome.html and www.ojp.usdoj.gov/ovc/assist/nvaa2002/chapter13.html.

[40] Ibid.

[41] Ibid.

[42] Sobieski, Regina. (1994). "MADD Study Finds Most Victims Satisfied with Law Enforcement." In *MADDVOCATE.* Irving, TX: Mothers Against Drunk Driving. Cited by Stephanie Frogge and Janice Lord (June 2002; updated December 2004), "Chapter 13: Drunk Driving." In Anne Seymour, Morna Murray, Jane Sigmon, Melissa Hook, Christine Edmunds, Mario Gaboury, and Grace Coleman. (eds.). (June 2002; updated July 2005). *National Victim Assistance Academy Textbook.* Washington, DC: U.S. Department of Justice, Office of Justice Programs, Office for Victims of Crime. Retrieved August 23, 2006, from www.ojp.usdoj.gov/ovc/assist/nvaa2002/welcome.html and www.ojp.usdoj.gov/ovc/assist/nvaa2002/chapter13.html.

[43] Miko, Francis T. (updated July 2006). "Summary." In *Trafficking in Persons: The U.S. and International Response.* (Congressional Research Service Report for Congress). Washington, DC: U.S. Congress, The Library of Congress, Congressional Research Service, Order Code RL30545. Retrieved April 16, 2007, from www.fpc.state.gov/documents/organization/70330.pdf.

[44] Ibid.

[45] Office to Monitor and Combat Trafficking in Persons. (June 2006). "Introduction." In *Trafficking in Persons Report.* Washington, DC: U.S. Department of State, Under Secretary for Democracy and Global Affairs, Office to Monitor and Combat Trafficking in Persons. Retrieved November 27, 2007, from www.state.gov/g/tip/rls/tiprpt/2006/65983.htm and www.state.gov/documents/organization/66086.pdf.

[46] Ibid.

[47] Ibid.

[48] Miko, Francis T. (updated July 2006).

[49] Office to Monitor and Combat Trafficking in Persons. (June 2006).

[50] Human Smuggling and Trafficking Center. (January 2005). *Fact Sheet: Distinctions Between Human Smuggling and Human Trafficking.* Washington, DC: U.S. Department of State, Under Secretary for Political Affairs, Bureau for International Narcotics and Law Enforcement Affairs, Human Smuggling and Trafficking Center. Retrieved April 16, 2007, from www.state.gov/documents/organization/49875.pdf

First Response to Victims of Crime

For copies of this guidebook and/or additional information,
please contact

OVC Resource Center
P.O. Box 6000
Rockville, MD 20849–6000
Telephone: 1–800–851–3420 or 301–519–5500
(TTY 1–877–712–9279)
www.ncjrs.gov

Or order OVC publications online at *www.ncjrs.gov/App/Publications/AlphaList.aspx*.
Submit your questions to Ask OVC at *http://ovc.ncjrs.gov/askovc*.
Send your feedback on this service via *www.ncjrs.gov/App/Feedback.aspx*.

Refer to publication number NCJ 231171.

For information on training and technical
assistance available from OVC, please contact

OVC Training and Technical Assistance Center
9300 Lee Highway
Fairfax, VA 22031
Telephone: 1–866–OVC–TTAC (1–866–682–8822)
(TTY 1–866–682–8880)
www.ovcttac.gov